Texts and Monographs in Physics

Springer
Berlin
Heidelberg
New York
Barcelona
Budapest
Hong Kong
London
Milan
Paris
Santa Clara
Singapore
Tokyo

Texts and Monographs in Physics

Series Editors: R. Balian W. Beiglböck H. Grosse E. H. Lieb
N. Reshetikhin H. Spohn W. Thirring

Georg Junker

Supersymmetric Methods in Quantum and Statistical Physics

With 36 Figures and 13 Tables

 Springer

Priv.-Doz. Dr. Georg Junker

Institut für Theoretische Physik
Universität Erlangen-Nürnberg
Staudtstrasse 7
D-91058 Erlangen, Germany

Editors

Roger Balian

CEA
Service de Physique Théorique de Saclay
F-91191 Gif-sur-Yvette, France

Wolf Beiglböck

Institut für Angewandte Mathematik
Universität Heidelberg
Im Neuenheimer Feld 294
D-69120 Heidelberg, Germany

Harald Grosse

Institut für Theoretische Physik
Universität Wien
Boltzmanngasse 5
A-1090 Wien, Austria

Elliott H. Lieb

Jadwin Hall
Princeton University, P. O. Box 708
Princeton, NJ 08544-0708, USA

Nicolai Reshetikhin

Department of Mathematics
University of California
Berkeley, CA 94720-3840, USA

Herbert Spohn

Theoretische Physik
Ludwig-Maximilians-Universität München
Theresienstrasse 37
D-80333 München, Germany

Walter Thirring

Institut für Theoretische Physik
Universität Wien
Boltzmanngasse 5
A-1090 Wien, Austria

Library of Congress Cataloging-in-Publication Data applied for.

Die Deutsche Bibliothek – CIP-Einheitsaufnahme

Junker, Georg:
Supersymmetric methods in quantum and statistical physics / Georg Junker. – Berlin ; Heidel-
berg ; New York ; Barcelona ; Budapest ; Hong Kong ; London ; Milan ; Paris ; Santa Clara ;
Singapore ; Tokyo : Springer, 1996
(Texts and monographs in physics)
Zugl.: Erlangen, Nürnberg, Univ., Habil.-Schr., 1995
ISBN-13:978-3-642-64742-0   e-ISBN-13:978-3-642-61194-0
DOI: 10.1007/978-3-642-61194-0

ISSN 0172-5998
ISBN-13:978-3-540-61591-0 Springer-Verlag Berlin Heidelberg New York

Typesetting: Camera-ready copy from the author using a Springer TeX macro package
Cover design: *design & production* GmbH, Heidelberg
SPIN: 10545662   55/3144-543210 - Printed on acid-free paper

Dedicated to the memory of

Asim Orhan Barut
(24 June 1926 – 5 December 1994)

But try to remember that a good man can never die.
The person of a man may go, but the best part of him stays.
It stays forever.

William Saroyan: The Human Comedy

Preface

The idea of supersymmetry was originally introduced in relativistic quantum field theories as a generalization of Poincaré symmetry. In 1976 Nicolai suggested an analogous generalization for non-relativistic quantum mechanics. With the one-dimensional model introduced by Witten in 1981, supersymmetry became a major tool in quantum mechanics and mathematical, statistical, and condensed-matter physics. Supersymmetry is also a successful concept in nuclear and atomic physics. An underlying supersymmetry of a given quantum-mechanical system can be utilized to analyze the properties of the system in an elegant and effective way. It is even possible to obtain exact results thanks to supersymmetry.

The purpose of this book is to give an introduction to supersymmetric quantum mechanics and review some of the recent developments of various supersymmetric methods in quantum and statistical physics. Thereby we will touch upon some topics related to mathematical and condensed-matter physics. A discussion of supersymmetry in atomic and nuclear physics is omitted. However, the reader will find some references in Chap. 9. Similarly, supersymmetric field theories and supergravity are not considered in this book. In fact, there exist already many excellent textbooks and monographs on these topics. A list may be found in Chap. 9. Yet, it is hoped that this book may be useful in preparing a footing for a study of supersymmetric theories in atomic, nuclear, and particle physics.

The plan of the book is as follows. Chapter 1 starts with a brief historical review of symmetry and supersymmetry.

In Chap. 2 the basic definitions for supersymmetric quantum mechanics are given and fundamental properties following from these definitions are discussed. In particular, we study the properties of $N = 2$ supersymmetric quantum systems. It is shown that the existence of the Witten-parity operator naturally leads to a grading of the Hilbert space. The supersymmetry transformation and the Witten index are introduced. Ground-state properties of a supersymmetric quantum system are also discussed.

Chapter 3 extensively discusses Witten's model, which in essence consists of a pair of standard one-dimensional Hamiltonians. This model is uniquely characterized by a scalar function which is often called the supersymmetric (SUSY) potential. Properties of this SUSY potential and the corresponding supersymmetric Witten model are studied. Many examples of SUSY potentials leading to well-known quantum problems in one dimension are presented.

Supersymmetric classical mechanics is introduced and studied in Chap. 4. Here we consider the pseudoclassical version of Witten's model. A pseudoclassical system possesses, in addition to the usual degrees of freedom, Grassmann-valued degrees of freedom. The equations of motion are explicitly solved and the properties of the solutions are discussed. Here emphasis is put on the so-called fermionic phase. Quantization of this pseudoclassical system is achieved via the canonical as well as the path-integral approach.

Exact solutions of quantum mechanical eigenvalue problems are covered in Chap. 5. An additional property of the SUSY potential called shape-invariance allows an explicit derivation of the discrete spectrum of the Hamiltonian in question and the corresponding eigenstates. The connection with Schrödinger's factorization method is shown.

As an approximation method we consider in Chap. 6 the quasi-classical path-integral approach to the Witten model. This approach yields novel quasi-classical quantization conditions having some remarkable properties. These are discussed in detail and are compared with the usual WKB approximation method.

An application in statistical physics is studied in Chap. 7. The super-symmetric structure of the Fokker–Planck and that of the Langevin equation are shown to be closely connected to the quantum mechanical and the pseudoclassical Witten model, respectively. Implications of supersymmetry are discussed. Supersymmetric approximation methods for decay rates are also presented.

In Chap. 8 we consider the supersymmetry of Pauli's Hamiltonian and discuss the paramagnetism of a two- and of a three-dimensional non-interacting electron gas, and the paramagnetic conjecture. Supersymmetry in Dirac's equation is briefly discussed and its application to semiconductor heterojunctions is also given.

The selection of the topics covered in this book is clearly subjective and results from my own interests. Chapter 9 is aimed at giving a broad overview of topics in theoretical physics where supersymmetric methods became and still are important tools.

The overview given in Chap. 9 is not meant to be complete. Similarly, the list of references, though rather extensive, is certainly incomplete. I apologize to all those who have been left out due to my ignorance.

This book originated in a course of lectures given at the University of Erlangen-Nürnberg during the winter term 1993/94. My view of supersymmetric quantum mechanics has been developed during many years of collaboration with Akira Inomata. Discussions with him, either personally or via e-mail, are always enjoyable and clarifying. His comments and advice have been very valuable for this book. Similar thanks go to Hajo Leschke. This book benefitted a lot from his clarifying comments and suggestions. Also several other people have read parts of the manuscript and made many valuable comments. I am especially indebted to Wolf Beiglböck, Helmut Fink, Werner Fischer, Alfred Hüller, John R. Klauder, Stephan Matthiesen, Peter Müller, and last but not least Siegfried Wonneberger for his endurance in reading the complete manuscript.

The collaboration with Akira Inomata on supersymmetric quantum mechanics has been supported by the Deutsche Forschungsgemeinschaft. This support is gratefully acknowledged. I am also grateful to Urda Beiglböck and Frank Holzwarth for their assistance with the Springer LaTeX macro package CLMono01.

Finally I am greatly indebted to my wife Karin and my children Julia and Michael for their moral support and patience.

Erlangen, June 1996   *Georg Junker*

Contents

1. Introduction

"Dass der Raum, als Ort für Puncte aufgefasst, nur drei Dimensionen hat, braucht vom mathematischen Standpuncte aus nicht discutirt zu werden; ebenso wenig kann man aber vom mathematischen Standpuncte aus Jemanden hindern, zu behaupten, der Raum habe eigentlich vier, oder unbegränzt viele Dimensionen, wir seien aber nur im Stande, drei wahrzunehmen."[1]

Felix Klein (1849–1925)

Symmetry plays important roles in theoretical physics. There are numerous kinds of symmetry in nature. Some are visible and some are hidden. Some are static and some are dynamic. Some belong to simple individual systems and some may be seen in the collective behavior of many systems. Often, various symmetries reveal themselves at the same time and complicate the appearance of a physical phenomenon. Mathematically, symmetry is handled by group theory; symmetry of a physical system is seen as an invariance under a group action. The visual symmetry of a crystal may be described by a discrete group. Symmetry of a system in motion is represented by a continuous group. The mirror-image symmetry of a system can be described by a discrete group of reflections and uniformity of time flow may be seen as a consequence of invariance under a continuous group of time translations.

In 1872, Felix Klein [Kl1872], in his inaugural lecture at Erlangen, commonly known as the Erlangen program, made an observation that geometry of space is associated with a mathematical group. According to Klein, the Euclidean space, for instance, is characterized by its transformation groups which consist of rotations, translations, and reflections. In 1918, Emmy Noether [Noe18] put forth another important theorem that if a system is invariant under a continuous group of n parameters and satisfies the equations of motion, there exist n constants of motion. As a result of Noether's theorem, we see that if a system moves freely in D-dimensional Euclidean space then the D-component momentum and $(D^2 - D)/2$-component angular momentum must

---

[1] F. Klein [Kl1872] p.42.

be conserved in association with the translation group of D parameters and the rotation group of $(D^2 - D)/2$ parameters, respectively. The symmetry considerations are useful not only in the study of classical systems but also in understanding quantum phenomena. Soon after the development of quantum mechanics the symmetry methods were found to be powerful in analyzing quantum spectra [Wey31, Wig31, Wae32].

In modern physics we recognize two kinds of symmetry; *external symmetry* and *internal symmetry*. In classical physics, we are concerned with external symmetries. In understanding the detailed structure of an atomic spectrum, it became necessary to introduce the concept of quantized spin. Since it is difficult to ascribe the spin concept associated with a point particle to the classical spinning of an extended body, the spin in non-relativistic quantum mechanics was understood as an internal degree of freedom. As quantum mechanics was applied to nuclear physics and high-energy physics, numerous additional internal degrees of freedom and their associated internal symmetries were introduced. There have been attempts to unify external symmetries and internal symmetries. However, such a grand unification has not been fully achieved so far.

In Minkowskian space-time, where quantum field theory is formulated, the maximal symmetry group is the Poincaré (or inhomogeneous Lorentz) group. It contains the homogeneous Lorentz group as a subgroup which allows for a classification of fundamental particles by their spin [Wig39]. Since the allowed values of spin are integers or half-integers, as long as our physical world is Minkowskian, fundamental particles must be either bosons or fermions. Thus, in the relativistic formulation, the spin is no longer associated with an internal symmetry but with a manifestation of the external space-time symmetry. Internal degrees of freedom such as isospin, baryon number, color, strangeness, charm, etc. are still associated with internal symmetries.

In 1967 Coleman and Mandula [CoMa67] investigated all possible symmetries of the scattering matrix in relativistic quantum field theory. Their result is as follows.[2] If one restricts the set of continuous symmetries to those generated by a *Lie algebra*, then the set of all possible generators consists only of the angular momentum Lorentz tensor $M_{\mu\nu}$, the momentum Lorentz vector P_μ, and Lorentz scalars. While $M_{\mu\nu}$ and P_μ generate the Poincaré group, additional symmetries must be generated by Lorentz scalars, which are indeed internal symmetries. However, the restriction to Lie symmetries made in the work of Coleman and Mandula has no *a priori* grounds.

---

[2] They considered field theory in more than two space-time dimensions with a finite number of massive one-particle states and a non-vanishing scattering amplitude.

In 1968, Miyazawa [Miy68] suggested a possible unification scheme for mesons and baryons based on a *superalgebra*. While a Lie algebra consists only of commutators, a superalgebra is closed with respect to commutators and anticommutators.[3] Gol'fand and Likhtman [GoLi71] were the first to embed the Poincaré algebra into a superalgebra. The supersymmetric field theory formulated in 1972 by Volkov and Akulov [VoAk72] was not renormalizable. In 1974, Wess and Zumino [WeZu74a, WeZu74b] succeeded in formulating a renormalizable supersymmetric field theory. Subsequently, Haag, Lopuszánski, and Sohnius [HaLoSo75] constructed "all possible generators of supersymmetries of the S-matrix". They found that within the framework of superalgebras, besides the generators of the Poincaré group and possible Lorentz scalars, there are spinor operators Q_α allowed. In the Weyl representation where the *supercharges* Q_α ($\alpha \in \{1,2\}$) are given by two components of a left-handed Weyl spinor, the superalgebra reads (without the Poincaré subalgebra and extra Lorentz scalars)

$$[Q_\alpha, M_{\mu\nu}] = \tfrac{1}{2} (\sigma_{\mu\nu})_\alpha^{\ \beta} Q_\beta, \qquad [Q_\alpha, P_\mu] = 0,$$
$$\{Q_\alpha, Q_\beta\} = 0, \qquad \{Q_\alpha, Q_\beta^\dagger\} = 2 (\sigma^\mu)_{\alpha\beta} P_\mu. \tag{1.1}$$

Here $[A, B] := AB - BA$ and $\{A, B\} := AB + BA$ denote the commutator and anticommutator, respectively. Furthermore,

$$\sigma^\mu := (\mathbf{1}, \boldsymbol{\sigma}) \equiv (\mathbf{1}, \sigma_1, \sigma_2, \sigma_3), \qquad \sigma^{\mu\nu} := \tfrac{1}{2} [\sigma^\mu, \sigma^\nu], \tag{1.2}$$

with Pauli matrices

$$\sigma_1 := \begin{pmatrix} 0 & 1 \\ 1 & 0 \end{pmatrix}, \qquad \sigma_2 := \begin{pmatrix} 0 & -i \\ i & 0 \end{pmatrix}, \qquad \sigma_3 := \begin{pmatrix} 1 & 0 \\ 0 & -1 \end{pmatrix}. \tag{1.3}$$

From this superalgebra we notice that

$$H \equiv P_0 = \tfrac{1}{4} \sum_\alpha \{Q_\alpha, Q_\alpha^\dagger\}, \qquad Q_\alpha^2 = 0. \tag{1.4}$$

From the first commutator of (1.1), it is also obvious that the supercharge operators change the eigenvalues of the third component M_{12} of the angular momentum operator by $\tfrac{1}{2}$. Therefore, each of the supercharge operators converts a bosonic state to a fermionic state and a fermionic state to a bosonic state, and becomes a generator of the so-called *supersymmetry transformations* (see Fig. 1.1). In a nucleus, the proton and the neutron are constantly transmuted into each other, so that they are not physically distinguishable. It is more appropriate to consider them as two possible states of a single nucleon, forming an iso-spinor. The indistinguishability of the proton and the neutron

---

[3] For details see, for example, the books by Scheunert [Sche79] and Cornwell [Cor89].

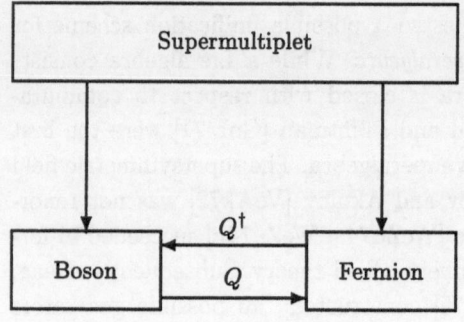

Fig. 1.1. The SUSY transformations in relativistic quantum field theories generated by supercharges $Q, Q^\dagger$.

in a nucleus lead Heisenberg to the idea of iso-symmetry. In much the same sense, the idea of supersymmetry assumes that there may be environments where bosons and fermions become indistinguishable. They are to be viewed as members of a single supermultiplet. In order for a physical system to have supersymmetry (SUSY), the ground state (vacuum) $|0\rangle$ of the system should be invariant under any SUSY transformation. This means that if SUSY is a good symmetry

$$Q_\alpha|0\rangle = 0 \quad \text{and} \quad Q_\alpha^\dagger|0\rangle = 0, \quad \text{for all } \alpha . \tag{1.5}$$

If SUSY is broken, then we have

$$Q_\alpha|0\rangle \neq 0 \quad \text{and/or} \quad Q_\alpha^\dagger|0\rangle \neq 0, \quad \text{for at least one } \alpha. \tag{1.6}$$

As has been mentioned earlier, SUSY was originally introduced to physics in search of a possible non-trivial unification of space-time and internal symmetries within four-dimensional relativistic quantum field theory. However, application of the SUSY idea is not limited to high-energy particle physics. SUSY has been successfully applied to other areas of theoretical physics such as nuclear, atomic, solid-state, and statistical physics [KoCa85]. Even for mathematical aspects of theoretical physics, it has been found to be a useful concept [KoCa85]. As will be shown in later chapters, SUSY has become a powerful tool in non-relativistic quantum mechanics. The first SUSY application at the non-relativistic level was done in 1976 by Nicolai [Nic76]. He utilized SUSY to construct spin models in statistical physics.

Although the SUSY idea is fascinating, we know that a boson and a fermion are quite clearly separate objects. In our surrounding environments, we can find no phenomenon in which a boson is converted into a fermion. In other words, as far as we see, SUSY is not a good symmetry. SUSY is in fact broken in our physical world. Yet, if we have a faith in SUSY, then it is reasonable to assume that SUSY has spontaneously been broken at some point in time and temperature. In 1981, Witten [Wit81] introduced SUSY quantum

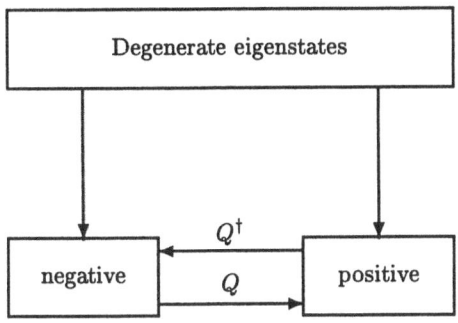

Fig. 1.2. The SUSY transformation of SUSY quantum mechanics transforms between states of positive and negative Witten parity.

mechanics, based on the simplest superalgebra, in order to provide a simple non-relativistic model for the spontaneous SUSY breaking mechanism. Witten's formulation of non-relativistic SUSY quantum mechanics attracted considerable attention in the last decade and is still serving as a useful tool in quantum physics [GeKr85, LaRoBa90, CoKhSu95]. Some recent textbooks on quantum mechanics [deLaRa91, Schw95] include Witten's formulation as a modern alternative to the factorization method of Schrödinger, Infeld, and Hull. The SUSY approach also serves as an important tool for analyzing mathematical structures of the Schrödinger equation and the Dirac equation [CyFrKiSi87, Gro91, Tha92]. It should also be mentioned that the recent work of Seiberg and Witten [SeWi94a, SeWi94b] again stimulated intensive research activities on four-dimensional supersymmetric gauge theories. Actually, Seiberg and Witten obtained, by means of electric-magnetic duality, exact information about the vacuum structure of $N = 2$ supersymmetric Yang–Mills theories. An overview on these results is given in [Sei95, InSe95].

Despite the fact that SUSY quantum mechanics is indeed the (0+1)-dimensional limit of SUSY quantum field theory, it is rather independent of the latter. SUSY in SUSY quantum mechanics is not the original supersymmetry relating bosons and fermions. The supercharges of SUSY quantum mechanics do not generate transformations between bosons and fermions. They generate transformations between two orthogonal eigenstates of a given Hamiltonian with the same degenerate eigenvalue (see Fig. 1.2). These two orthogonal states are eigenstates of the so-called Witten parity operator (see Chap. 2) with eigenvalues +1 and −1. Although the degeneracy arises from the fact that the supercharge operators change the eigenvalues by $\frac{1}{2}$, the Witten parity states in SUSY quantum mechanics may be taken in general as independent from the real spin states. Some of the examples we shall discuss later on are, indeed, models of a particle carrying a physical spin-$\frac{1}{2}$ degree of freedom. In order to avoid a possible confusion, we shall refer to those states that are independent of spin as the Witten-parity states.

2. Supersymmetric Quantum Mechanics

In this chapter we will start with the definition of the so-called N-extended super-symmetric quantum mechanics. This approach, actually for $N = 2$, has first been formulated by Nicolai [Nic76] in his search for supersymmetry in non-relativistic quantum systems related to models of statistical physics. Witten's [Wit81, Wit82a] approach has been for a general $N \geq 1$ but, for $N \geq 2$, has the same features as Nicolai's formulation. Here we will present the definition due to Witten in order to include the $N = 1$ case, which has some interesting applications in the theory of a three-dimensional electron gas in a magnetic field. However, only for $N \geq 2$ we will find the typical properties of supersymmetric quantum mechanics. In particular, $N \geq 2$ allows the introduction of the so-called Witten parity and thus implies a natural grading of the Hilbert space into two subspaces.

2.1 Definition of SUSY Quantum Mechanics

Let us suppose that we have a quantum system which is characterized by a Hamiltonian H acting on some Hilbert space $\mathcal{H}$. Furthermore, we will postulate the existence of N self-adjoint operators $Q_i = Q_i^\dagger$, $i = 1, 2, \ldots, N$, which also act on $\mathcal{H}$. With this setup we spell out the following definition.

Definition 2.1.1. *A quantum system, which is characterized by the set* $\{H, Q_1, \ldots, Q_N; \mathcal{H}\}$, *is called* supersymmetric *if the following anticommu-tation relation is valid for all* $i, j = 1, 2, \ldots, N$:

$$\boxed{\{Q_i, Q_j\} = H\delta_{ij},}$$

(2.1)

where δ_{ij} *denotes Kronecker's delta symbol. The self-adjoint operators* Q_i *are called* supercharges *and the Hamiltonian* H *is called* SUSY Hamiltonian. *The symmetry characterized by the superalgebra (2.1) is called* N-extended supersymmetry.

Let us remark that sometimes SUSY quantum mechanics is defined with non-selfadjoint supercharges. For example, for $N = 2M$ one can define the "complex" supercharges $\tilde{Q}_i := (Q_{2i} + iQ_{2i+1})/\sqrt{2}$. Because of this, an $N = 2M$ extended supersymmetry, in the above sense, is sometimes called an M-extended SUSY. We will exclusively use the notion of the above definition.

The above superalgebra certainly puts restrictions on the SUSY Hamiltonian H. First we notice that from the fundamental anticommutation relation (2.1) immediately follows

$$H = 2Q_1^2 = 2Q_2^2 = \cdots = 2Q_N^2 = \frac{2}{N}\sum_{i=1}^{N} Q_i^2, \tag{2.2}$$

which may be compared with (1.4). The supercharges of a supersymmetric quantum system are square roots of the SUSY Hamiltonian. A direct consequence of the relation $H = 2Q_i^2$ is

$$[H, Q_i] = 0 \qquad \text{for all} \qquad i = 1, 2, \ldots, N. \tag{2.3}$$

Therefore, the supercharges are constants of motion if they do not depend explicitly on time, $\partial Q_i/\partial t = 0$.

As a second consequence of (2.2) we note that the Hamiltonian does not have negative eigenvalues. In other words, the ground-state energy[1]

$$E_0 := \inf \operatorname{spec} H \tag{2.4}$$

of a supersymmetric quantum system is non-negative

$$E_0 \geq 0. \tag{2.5}$$

In analogy to supersymmetric field theories, see eq. (1.5), we introduce the notion of good and broken SUSY.

Definition 2.1.2. *A supersymmetric quantum system* $\{H, Q_1, \ldots, Q_N; \mathcal{H}\}$ *is said to have a good SUSY if its ground-state energy vanishes, that is,* $E_0 = 0$. *For a strictly positive ground-state energy* $E_0 > 0$ *SUSY is said to be broken.*

The analogy to (1.5) becomes visible through the following

Proposition 2.1.1. *For good SUSY all ground states* $|\psi_0^j\rangle$ *(j enumerates a possible degeneracy of the ground-state energy* $E_0 = 0$*) are annihilated by all supercharges:*

$$Q_i|\psi_0^j\rangle = 0 \qquad \text{for all } i \text{ and } j. \tag{2.6}$$

If SUSY is broken $(E_0 > 0)$ *there exists at least one pair* (i, j) *for which*

$$Q_i|\psi_0^j\rangle \neq 0. \tag{2.7}$$

Proof. Obviously, $H|\psi_0^j\rangle = E_0|\psi_0^j\rangle$ implies by (2.2) $\sum_{i=1}^{N} \||Q_i|\psi_0^j\rangle\|^2 = E_0 N/2$. Hence, $E_0 = 0$ implies (2.6) and $E_0 > 0$ implies (2.7), respectively.

---

[1] The symbol spec H denotes the spectrum (set of all eigenvalues) of the operator H.

It should be pointed out that supersymmetry imposes even more and stronger constraints on quantum systems. A rather general construction procedure for quantum systems with a SUSY structure as defined above has been given by de Crombrugghe and Rittenberg [CrRi83]. Below we will present some examples of supersymmetric quantum systems for $N = 1, 2$ and arbitrary positive integer values of N, respectively.

2.1.1 The Pauli Hamiltonian ($N = 1$)

As a first example we will mention a system which is characterized by the Pauli Hamiltonian for a spin-$\frac{1}{2}$ particle in an external magnetic field. The mass of this particle is denoted by m, its charge by e and the external magnetic field by $B(r,t) := \nabla \times A(r,t)$. Here the real-valued function $A : \mathbb{R}^3 \times \mathbb{R} \mapsto \mathbb{R}^3$ denotes the vector potential. Furthermore, c and $\hbar$ denote the speed of light and Planck's constant (divided by 2π), respectively. Because of the presence of spin the Hilbert space is given as the tensor product $\mathcal{H} := L^2(\mathbb{R}^3) \otimes \mathbb{C}^2$. Here $L^2(\mathbb{R}^3)$ is the infinite-dimensional Hilbert space of Lebesgue square-integrable complex-valued functions on the three-dimensional Euclidean space $\mathbb{R}^3$ and $\mathbb{C}^2$ is the two-dimensional Hilbert space equipped with the standard scalar product.

Let us define this supersymmetric system by introducing [CrRi83] the self-adjoint supercharge ($N = 1$):

$$Q_1 := \frac{1}{\sqrt{4m}} \left(p - \frac{e}{c} A \right) \cdot \sigma. \tag{2.8}$$

Due to the SUSY requirement (2.2) the Hamiltonian is necessarily given by $H_P := 2Q_1^2$. An explicit calculation leads to

$$H_P = \frac{1}{2m} \left(p - \frac{e}{c} A \right)^2 - \frac{e\hbar}{2mc} B \cdot \sigma, \tag{2.9}$$

and coincides with the well-known Pauli Hamiltonian for a spin-$\frac{1}{2}$ particle with gyromagnetic factor $g = 2$. It is an amusing and interesting observation [CrRi83] that supersymmetry suggests $g = 2$.[2] Without supersymmetry this usually follows from the relativistic covariant Dirac Hamiltonian only. In fact, relativity and supersymmetry are closely related [Cas76a]. To be more explicit, the supercharge (2.8) is related to the Dirac Hamiltonian (in the standard Pauli-Dirac representation) for the same point mass in the same magnetic field B:[3]

---

[2] In Sect. 8.1 we will show that supersymmetry also allows for a gyromagnetic factor $g = -2$.

[3] For the definition of a supersymmetric Dirac Hamiltonian see Sect. 8.3.

$$H_D := \begin{pmatrix} mc^2 & (cp - eA) \cdot \sigma \\ (cp - eA) \cdot \sigma & -mc^2 \end{pmatrix}$$

$$= \begin{pmatrix} mc^2 & 2\sqrt{mc^2}\,Q_1 \\ 2\sqrt{mc^2}\,Q_1 & -mc^2 \end{pmatrix}. \tag{2.10}$$

Let us also mention that the last expression for H_D immediately leads to a known [Fey61] relation between the Pauli and Dirac Hamiltonian:

$$H_D^2 = (mc^2)^2 \begin{pmatrix} 1 + 2H_P/mc^2 & 0 \\ 0 & 1 + 2H_P/mc^2 \end{pmatrix} \tag{2.11}$$

Obviously, the supercharge (2.8) commutes with H_P as well as with H_D and, therefore, for $\partial A/\partial t = 0$ it is a constant of motion [Fey61].

Finally, let us note that with $|\psi_E\rangle$ being an eigenstate of H_P with eigenvalue $E > 0$ the SUSY transformed state $|\psi_E'\rangle := \sqrt{2/E}\,Q_1|\psi_E\rangle$ is also an eigenstate of H_P with the same eigenvalue E. For a further discussion of supersymmetry in the Pauli and Dirac Hamiltonian system see Chap. 8.

2.1.2 Witten's SUSY Quantum Mechanics ($N = 2$)

The most popular model of an $N = 2$ SUSY quantum system has been introduced by Witten [Wit81, Wit82a]. This model describes a Cartesian degree of freedom which carries an additional internal spin-$\frac{1}{2}$-like degree of freedom. Hence, the Hilbert space is given as $\mathcal{H} := L^2(\mathbb{R}) \otimes \mathbb{C}^2$. The two supercharges have been defined by Witten as follows:

$$Q_1 := \frac{1}{\sqrt{2}} \left(\frac{p}{\sqrt{2m}} \otimes \sigma_1 + \Phi(x) \otimes \sigma_2 \right),$$

$$Q_2 := \frac{1}{\sqrt{2}} \left(\frac{p}{\sqrt{2m}} \otimes \sigma_2 - \Phi(x) \otimes \sigma_1 \right), \tag{2.12}$$

where Φ is a real-valued function, $\Phi : \mathbb{R} \to \mathbb{R}$, which, for convenience, is assumed to be continuously differentiable. Note that Φ is customarily called *SUSY potential*. This should not be confused with the superpotential usually introduced in supersymmetric quantum field theories. In fact, the SUSY potential Φ may be considered as the derivative of the superpotential.

The SUSY Hamiltonian is necessarily given by $H := 2Q_1^2 = 2Q_2^2$ and has the explicit form

$$H = \left(\frac{p^2}{2m} + \Phi^2(x) \right) \otimes 1 + \frac{\hbar}{\sqrt{2m}} \Phi'(x) \otimes \sigma_3, \tag{2.13}$$

where the prime denotes differentiation with respect to the argument, that is, $\Phi' := \frac{d\Phi}{dx}$. In the eigenbasis of σ_3 the Hamiltonian becomes diagonal in the $\mathbb{C}^2$-space,

$$H = \begin{pmatrix} H_+ & 0 \\ 0 & H_- \end{pmatrix}, \tag{2.14}$$

where

$$H_\pm := \frac{p^2}{2m} + \Phi^2(x) \pm \frac{\hbar}{\sqrt{2m}} \Phi'(x) \tag{2.15}$$

are standard Schrödinger operators acting on states in $L^2(\mathbb{R})$.

This *Witten model* is the simplest one which shows all typical features of supersymmetric quantum mechanics. We will discuss this model extensively in the next chapter.

2.1.3 Nicolai's Supersymmetric Harmonic-Oscillator Chain

As a last example we will mention the supersymmetric harmonic-oscillator chain introduced by Nicolai in 1976 [Nic76]. This model consists of a one-dimensional lattice with N sites. Each of these sites may be occupied by bosonic and fermionic degrees of freedom. The bosonic degrees of freedom are characterized by bosonic creation and annihilation operators $a_i^\dagger$ and a_i, $i = 1, 2, \ldots N$, acting on $\otimes_{i=1}^N L^2(\mathbb{R})$ and obeying the commutation relations

$$[a_i, a_j^\dagger] = \delta_{ij}, \qquad [a_i, a_k] = 0, \qquad [a_i^\dagger, a_k^\dagger] = 0. \tag{2.16}$$

The fermionic degrees of freedom are described by fermionic creation and annihilation operators $b_i^\dagger$ and b_i, which act on $\mathbb{C}^{2N} = \otimes_{i=1}^N \mathbb{C}^2$ and obey the anticommutation relations

$$\{b_i, b_j^\dagger\} = \delta_{ij}, \qquad \{b_i, b_k\} = 0, \qquad \{b_i^\dagger, b_k^\dagger\} = 0. \tag{2.17}$$

Let us define N supercharges by

$$Q_n := \sqrt{\frac{\hbar\omega}{2}} \sum_{i=1}^N \left(a_i^\dagger b_{i+n} + b_{i+n}^\dagger a_i \right), \quad n = 1, 2, \ldots, N, \quad \omega > 0, \tag{2.18}$$

where we used the cyclic boundary conditions $a_{i+N} = a_i$ and $b_{i+N} = b_i$. It is easily verified that the supercharges (2.18) obey the superalgebra (2.1). The Hamiltonian explicitly reads

$$H := 2Q_n^2 = \hbar\omega \sum_{i=1}^N \left(a_i^\dagger a_i + b_i^\dagger b_i \right) \tag{2.19}$$

and describes N non-interacting bosonic and fermionic harmonic oscillators. Using the explicit realization of the fermionic operators in terms of Pauli matrices, $b_i = \sigma_-^{(i)} := \frac{1}{2}\left(\sigma_1^{(i)} - i\sigma_2^{(i)} \right)$, this Hamiltonian reads

$$H = \hbar\omega \sum_{i=1}^N \left[(a_i^\dagger a_i + \tfrac{1}{2}) + \tfrac{1}{2} \sigma_3^{(i)} \right]. \tag{2.20}$$

Let us note that the supersymmetry of this example can be enlarged by introducing [CrRi83] complex supercharges $\widetilde{Q}_{ij} := a_i^\dagger b_j$. Finally, we remark that by generalizing the definition (2.18) it is possible to find supersymmetric spin-chain models with a non-trivial interaction [Nic76].

2.2 Properties of $N = 2$ SUSY Quantum Mechanics

So far the Definition 2.1.1. does not necessarily give rise to degenerated energy eigenvalues of the Hamiltonian H, which in turn allow for a construction of supercharges generating a SUSY transformation (cf. Fig. 1.2). In this section we are going to show that for $N = 2$, and therefore also for $N > 2$, one can define an additional operator, which we will call Witten-parity operator. Analyzing the properties of this operator we will find that only $N \geq 2$ in general implies a degeneracy of the eigenvalues of H and thus allows for the construction of the corresponding SUSY transformation which transforms the associate energy eigenstates into each other.

The $N = 2$ SUSY quantum mechanics consists of two supercharges Q_1, Q_2 and a Hamiltonian H which obey the following relations:

$$Q_1 Q_2 = -Q_2 Q_1, \qquad H = 2Q_1^2 = 2Q_2^2 = Q_1^2 + Q_2^2. \tag{2.21}$$

Let us introduce the complex supercharges

$$Q := \frac{1}{\sqrt{2}}\left(Q_1 + iQ_2\right), \qquad Q^\dagger = \frac{1}{\sqrt{2}}\left(Q_1 - iQ_2\right). \tag{2.22}$$

These operators together with the Hamiltonian H close the superalgebra

$$\boxed{Q^2 = 0 = \left(Q^\dagger\right)^2, \qquad \{Q, Q^\dagger\} = H.} \tag{2.23}$$

It should be remarked that the algebra (2.23) is the one which has been introduced by Nicolai [Nic76] in his approach to SUSY quantum mechanics and is clearly motivated by the algebra of supersymmetric quantum field theories (cf. eq. 1.4). Indeed, many authors restrict their definition of SUSY quantum mechanics to the case $N = 2$. This is, because this case is sufficiently general to establish all typical properties of SUSY quantum mechanics. Nevertheless, there are interesting physical systems with $N = 1$ and $N > 2$, too.

2.2.1 The Witten Parity

Let us now, in addition to the operators forming the superalgebra (2.23), postulate the existence of a self-adjoint operator W, which obeys the relations

$$\boxed{[W, H] = 0, \qquad \{W, Q\} = 0 = \{W, Q^\dagger\}, \qquad W^2 = 1.}$$ (2.24)

That is, W should commute with the Hamiltonian, anticommute with the complex supercharges and define a unitary involution on $\mathcal{H}$.

Definition 2.2.1. *A self-adjoint operator W which obeys relations (2.24) is called* Witten parity *or* Witten operator.

It should be noted that for $N = 2$ such an operator may always be found by setting

$$W := \frac{2}{H} QQ^\dagger - 1 = \frac{1}{iH}[Q_1, Q_2] = \frac{[Q, Q^\dagger]}{\{Q, Q^\dagger\}},$$ (2.25)

which is only well defined on the orthogonal complement of ker H. That is, on the subspace spanned by the eigenvectors of H with strictly positive energy eigenvalues. However, the above explicit form (2.25) of W often has a natural extension to ker H. For example, in Witten's model, where Q_1, Q_2 and H are given by (2.12) and (2.13), respectively, we find $[Q_1, Q_2] = iH\sigma_3$. Hence, according to (2.25) the Witten operator is given by σ_3 and is well defined on the full Hilbert space $\mathcal{H}$. If there does not exist such a natural extension one may define W on ker H, for example, by postulating that all zero modes of H are positive or negative states. See Definition 2.2.3. below. It should be remarked that such a problem may only occur for good SUSY.

Being a unitary involution the Witten operator takes only the eigenvalues ± 1 justifying the name parity. Instead of W, Witten [Wit82a] originally considered the operator $(-1)^{\mathcal{F}} = -W$ where $\mathcal{F}$ denotes the so-called "fermion-number operator" [SaHo82]. Actually, it is even possible to introduce (on the complement of ker H) a "fermionic" annihilation operator $b := Q^\dagger/\sqrt{H}$ obeying the anticommutation relation $\{b, b^\dagger\} = 1$. Hence, the "fermion-number" operator $\mathcal{F} := b^\dagger b = QQ^\dagger/H = \mathcal{F}^\dagger = \mathcal{F}^2$ obeys the algebra $[\mathcal{F}, H] = 0$, $[\mathcal{F}, Q] = Q$, $[\mathcal{F}, Q^\dagger] = -Q^\dagger$, and is related to the Witten parity by $W = 2\mathcal{F} - 1 = (-1)^{\mathcal{F}+1}$. The eigenspace of W with eigenvalue $+1$ ($\mathcal{F} = 1$) is sometimes called "fermionic" subspace. Whereas, the eigenspace of W corresponding to the eigenvalue -1 ($\mathcal{F} = 0$) is called "bosonic" subspace. As this notion might be confusing, SUSY quantum mechanics does not deal with the real boson-fermion symmetry, we call them subspaces of positive and negative (Witten) parity, respectively:

Definition 2.2.2. *Let $P^{\pm} := \frac{1}{2}(1 \pm W)$ be the orthogonal projection of $\mathcal{H}$ onto the eigenspace of the Witten operator with eigenvalue ± 1. The subspace*

$$\mathcal{H}^{\pm} := P^{\pm}\mathcal{H} = \{|\psi\rangle \in \mathcal{H} : W|\psi\rangle = \pm|\psi\rangle\} \tag{2.26}$$

is called space of positive $(\mathcal{H}^{+})$ and negative $(\mathcal{H}^{-})$ Witten parity, respectively.

It is natural to decompose the Hilbert space into these eigenspaces of W,

$$\mathcal{H} = \mathcal{H}^{+} \oplus \mathcal{H}^{-}, \tag{2.27}$$

and represent each linear operator acting on $\mathcal{H}$ by 2×2 matrices.[4] For example, the Witten parity W and the projectors $P^{\pm}$ read in this notation

$$W = \begin{pmatrix} 1 & 0 \\ 0 & -1 \end{pmatrix}, \qquad P^{+} = \begin{pmatrix} 1 & 0 \\ 0 & 0 \end{pmatrix}, \qquad P^{-} = \begin{pmatrix} 0 & 0 \\ 0 & 1 \end{pmatrix}. \tag{2.28}$$

Because of $Q^2 = 0$ and $\{Q, W\} = 0$ the complex supercharges are necessarily of the form[5]

$$Q = \begin{pmatrix} 0 & A \\ 0 & 0 \end{pmatrix}, \qquad Q^{\dagger} = \begin{pmatrix} 0 & 0 \\ A^{\dagger} & 0 \end{pmatrix}, \tag{2.29}$$

which imply

$$Q_1 = \frac{1}{\sqrt{2}} \begin{pmatrix} 0 & A \\ A^{\dagger} & 0 \end{pmatrix}, \qquad Q_2 = \frac{i}{\sqrt{2}} \begin{pmatrix} 0 & -A \\ A^{\dagger} & 0 \end{pmatrix}. \tag{2.30}$$

Here $A : \mathcal{H}^{-} \to \mathcal{H}^{+}$ denotes a generalized "annihilation" operator and $A^{\dagger} : \mathcal{H}^{+} \to \mathcal{H}^{-}$ is its adjoint, which may be understood as a generalized "creation" operator [CyFrKiSi87].

The SUSY Hamiltonian becomes diagonal in the representation (2.28),

$$H = \begin{pmatrix} AA^{\dagger} & 0 \\ 0 & A^{\dagger}A \end{pmatrix}. \tag{2.31}$$

Hence, for an $N = 2$ supersymmetric quantum system the total SUSY Hamiltonian H consists of two so-called *SUSY partner Hamiltonians*

$$H_{+} := AA^{\dagger} \geq 0, \qquad H_{-} := A^{\dagger}A \geq 0. \tag{2.32}$$

Let us mention that an arbitrary operator B acting on $\mathcal{H}$ can be decomposed into its diagonal (even) part B_{e} and its off-diagonal (odd) part B_{o}. That is, $B = B_{\mathrm{e}} + B_{\mathrm{o}}$ with

---

[4] Because of this grading of $\mathcal{H}$ the Witten operator is sometimes called *grading operator*.

[5] The second possibility $Q = \begin{pmatrix} 0 & 0 \\ A & 0 \end{pmatrix}$ is, in essence, equivalent to (2.29).

$$[W, B_e] = 0, \qquad \{W, B_o\} = 0. \tag{2.33}$$

In particular, the SUSY Hamiltonian H is an even operator, whereas the supercharges Q and $Q^\dagger$ are odd operators.

Finally, we note that for $N = 1$ there does not necessarily exist a Witten operator. For this reason, some authors [GrPi87, JaLeLe87, CyFrKiSi87, Tha92] include the existence of such a Witten operator in their definition of supersymmetric quantum systems.

2.2.2 SUSY Transformation

According to the Definition 2.2.2. we introduce the notion of positive and negative Witten-parity states.

Definition 2.2.3. *Eigenstates of $P^\pm$ are called* positive *and* negative *(Witten-) parity states, respectively. They are denoted by $|\psi^\pm\rangle$:*

$$P^\pm|\psi^\pm\rangle = \pm|\psi^\pm\rangle. \tag{2.34}$$

For simplicity we will call them also positive and negative states.

In the matrix notation of (2.28) they are represented in the form

$$|\psi^+\rangle = \begin{pmatrix} |\phi^+\rangle \\ 0 \end{pmatrix}, \qquad |\psi^-\rangle = \begin{pmatrix} 0 \\ |\phi^-\rangle \end{pmatrix}, \tag{2.35}$$

where $|\phi^\pm\rangle \in \mathcal{H}^\pm$.

It is obvious that the supercharges (2.29) generate a SUSY transformation in the sense that they map negative states into positive states and vice versa:

$$Q\mathcal{H}^- \subset \mathcal{H}^+, \qquad Q^\dagger\mathcal{H}^+ \subset \mathcal{H}^-. \tag{2.36}$$

This SUSY transformation can be made explicit for eigenstates of H. Note that $[W, H] = 0$ and therefore, the Hamiltonian and the Witten operator have simultaneous eigenstates:

Proposition 2.2.1. *To each positive (negative) eigenstate $|\psi_E^+\rangle$ ($|\psi_E^-\rangle$) of the Hamiltonian H with eigenvalue $E > 0$ there exists a negative (positive) eigenstate of H with the same eigenvalue. These eigenstates are related by the SUSY transformation*

$$\boxed{|\psi_E^-\rangle = \frac{1}{\sqrt{E}} Q^\dagger|\psi_E^+\rangle, \qquad |\psi_E^+\rangle = \frac{1}{\sqrt{E}} Q|\psi_E^-\rangle.} \tag{2.37}$$

Proof. Let, for example, $|\psi_E^-\rangle$ be a negative eigenstate of H, that is, $H|\psi_E^-\rangle = E|\psi_E^-\rangle$. Then, because of $[H, Q] = 0$, we have $HQ|\psi_E^-\rangle = QH|\psi_E^-\rangle = EQ|\psi_E^-\rangle \in \mathcal{H}^+$. Hence, the normalized vector $(1/\sqrt{E})Q|\psi_E^-\rangle$ is a positive eigenstate of H for the same eigenvalue $E > 0$.

Corollary 2.2.1 ([Dei78]). *The spectra of the two SUSY partner Hamiltonians H_+ and H_- are identical away from zero:*

$$\operatorname{spec}(H_+)\backslash\{0\} = \operatorname{spec}(H_-)\backslash\{0\}. \tag{2.38}$$

If $|\phi_E^\pm\rangle \in \mathcal{H}^\pm$ denotes an eigenstate of $H_\pm$ with eigenvalue $E > 0$, that is, $H_\pm|\phi_E^\pm\rangle = E|\phi_E^\pm\rangle$, then the corresponding SUSY transformation (2.37) reads

$$\boxed{|\phi_E^-\rangle = \frac{1}{\sqrt{E}} A^\dagger|\phi_E^+\rangle, \qquad |\phi_E^+\rangle = \frac{1}{\sqrt{E}} A|\phi_E^-\rangle.} \tag{2.39}$$

In other words, the strictly positive eigenvalues of the SUSY partner Hamiltonians $H_\pm$ coincide. Operators having the property (2.38) are said to be *essential iso-spectral* [Dei78, Schm78]. Hence, the Hamiltonian of $N = 2$ SUSY quantum mechanics consists of a pair of essential iso-spectral SUSY partner Hamiltonians. For more properties on essential iso-spectral operators we refer to [Dei78, Tha92]. We note that the above relations (2.37) and (2.39) are also valid for continuous spectra. For the proof, however, one needs resolvents [Dei78].

2.2.3 Ground-State Properties for Good SUSY

Until now we have considered eigenstates of the SUSY Hamiltonian H with strictly positive energies $E > 0$. For good SUSY, by definition, there exists (at least) one state in $\mathcal{H}$ with vanishing energy eigenvalue. Let us denote such a state by $|\psi_0\rangle$, that is $H|\psi_0\rangle = 0$. Necessarily, $|\psi_0\rangle$ is annihilated by the supercharges Q and $Q^\dagger$:

$$Q|\psi_0\rangle = 0, \qquad Q^\dagger|\psi_0\rangle = 0. \tag{2.40}$$

For a negative ground state $|\psi_0^-\rangle = \begin{pmatrix} 0 \\ |\phi_0^-\rangle \end{pmatrix}$ this implies

$$A|\phi_0^-\rangle = 0. \tag{2.41}$$

Whereas, a positive ground state $|\psi_0^+\rangle = \begin{pmatrix} |\phi_0^+\rangle \\ 0 \end{pmatrix}$ requires

$$A^\dagger|\phi_0^+\rangle = 0. \tag{2.42}$$

In general, the ground state $|\psi_0\rangle$ could be of positive or negative Witten parity. If the ground-state energy $E_0 = 0$ is degenerate even both types of states may occur. It should be noted, that in general, these states are not paired like those for strictly positive-energy eigenvalues. For typical spectra of a SUSY Hamiltonian in the case of good and broken SUSY see Fig. 2.1.

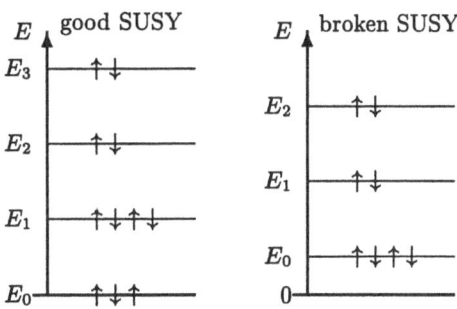

Fig. 2.1. Typical spectra for good ($E_0 = 0$) and broken ($E_0 > 0$) SUSY. Strictly positive-energy eigenvalues occur in positive-negative-parity pairs ($\uparrow\downarrow$).

2.2.4 The Witten Index

In the above we have seen that the ground-state properties of a SUSY quantum system may be of particular interest. In order to decide whether there are such *zero modes*, that is, states with zero energy, Witten [Wit82a] introduced the following quantity.

Definition 2.2.4. *Let us denote by $n_\pm$ the number of zero modes of $H_\pm$ in the subspace $\mathcal{H}^\pm$. For finite n_+ and n_- the quantity*

$$\boxed{\Delta := n_- - n_+} \tag{2.43}$$

is called the Witten *index.*

It is obvious that for $n_+ + n_- > 0$, by definition, SUSY is good. Hence, whenever the Witten index is a non-zero integer SUSY is good. However, if $\Delta = 0$ it is not clear whether SUSY is broken ($n_+ = 0 = n_-$) or not ($n_+ = n_- \neq 0$).

It should be remarked that the Witten index is related to the so-called Fredholm index [Kat84] of the annihilation operator A characterizing the supercharge Q:

$$\text{ind } A := \dim \ker A - \dim \ker A^\dagger = \dim \ker A^\dagger A - \dim \ker AA^\dagger, \tag{2.44}$$

where $\dim \ker A$ denotes the dimension of the space spanned by the linearly independent zero modes of the operator A (the kernel of A). Clearly, (2.44) makes only sense if both, $\dim \ker A$ and $\dim \ker A^\dagger$, are finite. Operators having this property are called Fredholm operators [Kat84]. The most remarkable property of the Fredholm index is its "topological invariance" [Kat84]. The Witten index obviously is related to the Fredholm index by

$$\Delta = \text{ind } A = \dim \ker H_- - \dim \ker H_+. \tag{2.45}$$

For non-Fredholm operators, that is, for systems where Definition 2.2.4 does not apply, it has been suggested [Cal78, Wei79, Wit82a, Jun95] to consider so-called regularized indices:

$$\bar{\Delta}(\beta) := \text{Tr}\left(-We^{-\beta H}\right) \tag{2.46}$$

$$= \text{Tr}_-\left(e^{-\beta A^\dagger A}\right) - \text{Tr}_+\left(e^{-\beta AA^\dagger}\right),$$

$$\hat{\Delta}(z) := \text{Tr}\left(-W\frac{z}{H-z}\right) \tag{2.47}$$

$$= \text{Tr}_-\left(\frac{z}{A^\dagger A - z}\right) - \text{Tr}_+\left(\frac{z}{AA^\dagger - z}\right),$$

$$\tilde{\Delta}(\varepsilon) := \text{Tr}\left(-W\Theta(\varepsilon - H)\right) \tag{2.48}$$

$$= \text{Tr}_-\left(\Theta(\varepsilon - A^\dagger A)\right) - \text{Tr}_+\left(\Theta(\varepsilon - AA^\dagger)\right),$$

where $\beta > 0$, $z < 0$ and $\varepsilon > 0$, respectively, $\text{Tr}_\pm(\cdot)$ stands for the trace in the subspaces $\mathcal{H}^\pm$, and Θ denotes the unit-step function, $\Theta(x) = 0$ for $x < 0$ and $\Theta(x) = 1$ for $x > 0$. The above indices are called *heat kernel regularized*, *resolvent kernel regularized* and *IDOS kernel regularized* index, respectively, for obvious reasons. Here IDOS stands for integrated density of states. The heat kernel regularized index has been introduced by Atiyah, Bott and Patodi [AtBoPa73] for an alternative proof of the Atiyah–Singer index theorem. It has also been used by Atiyah, Padoti and Singer [AtPaSi75] in their study of the spectral asymmetry of certain elliptic self-adjoint operators. For a derivation of the Atiyah–Singer index theorem based on supersymmetric quantum mechanics see [Alv83, FrWi84, MaZu86, Jar89].

For A being a Fredholm operator these regularized indices are independent of their argument and are identical to the Fredholm and Witten index:

$$\Delta = \bar{\Delta}(\beta) = \hat{\Delta}(z) = \tilde{\Delta}(\varepsilon) = \text{ind}\, A. \tag{2.49}$$

As already mentioned, for A being not a Fredholm operator the above definition for the Witten index cannot be used. In this case there are several alternative definitions available in the literature,

$$\Delta := \lim_{\beta \to \infty} \bar{\Delta}(\beta), \tag{2.50}$$

$$\Delta := \lim_{z \uparrow 0} \hat{\Delta}(z), \tag{2.51}$$

$$\Delta := \lim_{\varepsilon \downarrow 0} \tilde{\Delta}(\varepsilon), \tag{2.52}$$

whenever the quantity on the right-hand side exists.

The heat kernel and resolvent kernel regularized indices are well studied. See, for example, [Cal78, CeGi83, Hir83, NiWi84, AkCo84, BoBl84, JaLeLe87, BoGeGrSchwSi87, Nak90, Tha92]. The IDOS kernel regularized index has been shown to be related to the magnetization of a non-interacting electron gas in an arbitrary magnetic field [Jun95]. See also Sect. 8.2.

2.2.5 SUSY and Gauge Transformations

As a last point we will mention that the $N = 2$ superalgebra (2.21), respectively (2.23) is invariant under the "rotation"

$$\begin{pmatrix} Q_1' \\ Q_2' \end{pmatrix} := \begin{pmatrix} \cos\alpha & \sin\alpha \\ -\sin\alpha & \cos\alpha \end{pmatrix} \begin{pmatrix} Q_1 \\ Q_2 \end{pmatrix}, \qquad \alpha \in [0, 2\pi). \tag{2.53}$$

This invariance becomes explicit by noting that

$$Q_1'Q_2' = Q_1Q_2 = -Q_2Q_1 = -Q_2'Q_1', \qquad Q_1'^{\,2} = H/2 = Q_2'^{\,2}. \tag{2.54}$$

Similarly, for the complex supercharges one gets

$$Q' := e^{-i\alpha}Q, \qquad Q'^\dagger = e^{i\alpha}Q^\dagger, \tag{2.55}$$

which obviously obey the algebra (2.23).

It is interesting to note that this invariance of the superalgebra can be related to global gauge transformations, that is, changing the states in $\mathcal{H}^\pm$ by a constant phase:

$$|\psi'^\pm\rangle := e^{-i\beta_\pm}|\psi^\pm\rangle. \tag{2.56}$$

Note that for $\beta_+ \neq \beta_-$ the above global gauge transformations are different in the two subspaces $\mathcal{H}^+$ and $\mathcal{H}^-$, respectively. The SUSY relations (2.37) remain invariant if we perform the rotation (2.53) with a rotation angle given by $\alpha = \beta_+ - \beta_-$. Hence, the parameter α accounts for different global gauges in the two subspaces $\mathcal{H}^\pm$.

3. The Witten Model

In this chapter we will study a particular supersymmetric quantum model introduced by Witten in 1981 [Wit81]. Originally, this model has been designed to serve as a simple example for studying the SUSY-breaking mechanism in quantum field theories. However, in the last decade this model has found applications in many other areas of theoretical physics. Some of these applications will be discussed in this book.

3.1 Witten's Model and Its Modification

The Witten model which is an $N = 2$ extended supersymmetric quantum system is characterized be the two supercharges

$$
\begin{aligned}
Q_1 &:= \frac{1}{\sqrt{2}} \left(-\frac{p}{\sqrt{2m}} \otimes \sigma_2 + \Phi(x) \otimes \sigma_1 \right), \\
Q_2 &:= \frac{1}{\sqrt{2}} \left(\frac{p}{\sqrt{2m}} \otimes \sigma_1 + \Phi(x) \otimes \sigma_2 \right).
\end{aligned}
\tag{3.1}
$$

Note that this definition of the supercharges slightly differs from Witten's original one (2.12) by replacing $Q_1 \to Q_2$ and $Q_2 \to -Q_1$. Actually, this replacement corresponds to a rotation (2.53) with angle $\alpha = \pi/2$. As modification to Witten's setup we also restrict the configuration space of the point mass m to a one-dimensional subspace $\mathcal{M} \subseteq \mathbb{R}$. Here $\mathcal{M}$ either denotes the Euclidean line $\mathbb{R}$ (Witten's original model), the half line $\mathbb{R}^+ := [0, \infty)$ or a finite interval $[a, b] \subset \mathbb{R}$. Hence, the Hilbert space is given by $\mathcal{H} = L^2(\mathcal{M}) \otimes \mathbb{C}^2$. Let us note that for the latter two cases the supercharges (3.1) are in general symmetric but not self-adjoint. In addition one has to impose boundary condition for the wave functions at $x = 0$ and $x = a, b$, respectively [Ric78, ReSi80]. If not stated otherwise we will consider only the first case, that is, $\mathcal{M} = \mathbb{R}$. However, the results derived below are also valid for the other cases if we do not explicitly state the contrary. As before, we assume that the SUSY potential is a continuous differentiable function on $\mathcal{M}$, $\Phi : \mathcal{M} \to \mathbb{R}$.

In analogy to Sect. 2.2 we may introduce the complex supercharge

$$Q = \begin{pmatrix} 0 & A \\ 0 & 0 \end{pmatrix}, \qquad Q^\dagger = \begin{pmatrix} 0 & 0 \\ A^\dagger & 0 \end{pmatrix}, \tag{3.2}$$

where now the annihilation and creation operators are explicitly given by

$$A := \frac{ip}{\sqrt{2m}} + \Phi(x), \qquad A^\dagger = -\frac{ip}{\sqrt{2m}} + \Phi(x). \tag{3.3}$$

The SUSY Hamiltonian reads

$$H = \begin{pmatrix} H_+ & 0 \\ 0 & H_- \end{pmatrix}, \tag{3.4}$$

where the SUSY partner Hamiltonians are given by

$$H_\pm := \frac{p^2}{2m} + V_\pm(x) \tag{3.5}$$

with

$$V_\pm(x) := \Phi^2(x) \pm \frac{\hbar}{\sqrt{2m}} \Phi'(x). \tag{3.6}$$

Let us also mention that the commutator and anticommutator of the creation and annihilation operators (3.3) can be related to particular parts of the Hamiltonians $H_\pm = H_{\text{tree}} \pm H_{\text{loop}}$:

$$\begin{aligned} H_{\text{tree}} &:= \frac{p^2}{2m} + \Phi^2(x) = \frac{1}{2}\{A^\dagger, A\}, \\ H_{\text{loop}} &:= \frac{\hbar}{\sqrt{2m}}\Phi'(x) = \frac{1}{2}[A^\dagger, A]. \end{aligned} \tag{3.7}$$

The interpretation of the two parts H_{tree} and H_{loop} as "tree Hamiltonian" and "loop correction" (due to Grassmannian degrees of freedom) will be explained in Chap. 4.

3.2 Witten Parity and SUSY Transformation

It might be instructive to verify some of the general properties discussed in Sect. 2.2 for the Witten model. First, let us consider the Witten parity defined in (2.25). A simple computation shows that $[Q_1, Q_2] = iH\sigma_3$ and, hence, the Witten parity is indeed given by $W = \sigma_3$. Following the general treatment, we can grade the Hilbert space into eigenspaces of W. Obviously, the positive and negative Witten-parity subspaces are $\mathcal{H}^\pm = L^2(\mathcal{M})$. Here we note that for the case of $\mathcal{M} = \mathbb{R}^+$ and $\mathcal{M} = [a, b]$ the two subspaces $\mathcal{H}^\pm$ may have different boundary conditions. In many cases the positive and negative subspaces are identical, but there exist exceptions. Examples of such exceptions are given in (3.51) and (3.53) below.

Secondly, for completeness, we give the explicit form of the SUSY transformation (2.37), respectively, (2.39). If $|\phi_E^\pm\rangle$ denotes eigenstates of $H_\pm$ to the same eigenvalue E, the SUSY transformation explicitly reads

$$A|\phi_E^-\rangle = \left(\frac{ip}{\sqrt{2m}} + \Phi(x)\right)|\phi_E^-\rangle = \sqrt{E}|\phi_E^+\rangle,$$

$$A^\dagger|\phi_E^+\rangle = \left(-\frac{ip}{\sqrt{2m}} + \Phi(x)\right)|\phi_E^+\rangle = \sqrt{E}|\phi_E^-\rangle. \tag{3.8}$$

In the coordinate representation $\phi_E^\pm(x) := \langle x|\phi_E^\pm\rangle$ these relations take the form

$$\left(\pm\frac{\hbar}{\sqrt{2m}}\frac{\partial}{\partial x} + \Phi(x)\right)\phi_E^\pm(x) = \sqrt{E}\,\phi_E^\pm(x). \tag{3.9}$$

3.3 The SUSY Potential and Zero-Energy States

3.3.1 Ground State for Good SUSY

One of the important features of the Witten model is, that it allows for a rather explicit discussion of its ground-state properties. That is, one can determine from the shape of the SUSY potential whether SUSY is broken or not. In the good-SUSY case it is even possible to give the ground-state wave function explicitly.

For example, let us assume that SUSY is good ($E_0 = 0$) and the corresponding zero-energy eigenstate belongs to $\mathcal{H}^-$. That is, the ground state has negative Witten parity. This state necessarily satisfies eq. (2.41) which reads in the present model

$$\left(\frac{\hbar}{\sqrt{2m}}\frac{\partial}{\partial x} + \Phi(x)\right)\phi_0^-(x) = 0. \tag{3.10}$$

This first-order differential equation is easily integrated to

$$\phi_0^-(x) = C\exp\left\{-\frac{\sqrt{2m}}{\hbar}\int_{x_0}^x dz\,\Phi(z)\right\}, \tag{3.11}$$

where $x_0 \in \mathcal{M}$ is an arbitrary but fixed constant and $C = \phi_0^-(x_0)$ stands for a normalization constant. Introducing the so-called *superpotential*

$$U(x) := \frac{\sqrt{2m}}{\hbar}\int_{x_0}^x dz\,\Phi(z) \tag{3.12}$$

the ground-state wave function reads

$$\phi_0^-(x) = C \exp\{-U(x)\}. \tag{3.13}$$

Similarly, the assumption that the ground state belongs to the positive subspace $\mathcal{H}^+$ leads to

$$\phi_0^+(x) = \phi_0^+(x_0) \exp\{+U(x)\} \propto \frac{1}{\phi_0^-(x)}. \tag{3.14}$$

If SUSY is good, that is, there exists a zero-energy eigenstate for the SUSY Hamiltonian H, this state belongs either to $\mathcal{H}^-$ or $\mathcal{H}^+$. This is obvious, because if (3.13) is square integrable then (3.14) cannot be square integrable (and vice versa). In any case, the eigenvalue $E_0 = 0$ is not degenerate.

We also note that in case of good SUSY there is a close connection between the SUSY potential, the superpotential and the ground-state wave function [Goz83] expressed by the relations

$$U(x) = \pm \ln \frac{\phi_0^\pm(x)}{\phi_0^\pm(x_0)}, \qquad \Phi(x) = \pm \frac{\hbar}{\sqrt{2m}} \frac{(\phi_0^\pm)'(x)}{\phi_0^\pm(x)}, \tag{3.15}$$

where the upper sign is valid if the ground state belongs to $\mathcal{H}^+$ and the lower one if it belongs to $\mathcal{H}^-$, respectively.

3.3.2 An Additional Symmetry

As the reader may have realized, the two SUSY partner Hamiltonians $H_\pm$ differ only by an overall sign in the SUSY potential. In other words, changing the sign of Φ simply replaces H_+ by H_- and vice versa. Hence, the Witten model is invariant under the simultaneous replacements

$$\Phi \to \tilde{\Phi} := -\Phi, \qquad H_\pm \to \tilde{H}_\pm := H_\mp,$$

$$H \to \tilde{H} := \begin{pmatrix} \tilde{H}_+ & 0 \\ 0 & \tilde{H}_- \end{pmatrix}. \tag{3.16}$$

The above transformation implies

$$Q_1 \to \tilde{Q}_1 := \frac{1}{\sqrt{2}} \left(-\frac{p}{\sqrt{2m}} \otimes \sigma_2 - \Phi(x) \otimes \sigma_1 \right)$$

$$Q_2 \to \tilde{Q}_2 := \frac{1}{\sqrt{2}} \left(\frac{p}{\sqrt{2m}} \otimes \sigma_1 - \Phi(x) \otimes \sigma_2 \right) \tag{3.17}$$

and hence the superalgebra remains invariant, that is,

$$\{\tilde{Q}_i, \tilde{Q}_j\} = \tilde{H} \delta_{ij}. \tag{3.18}$$

However, we note that the transformation (3.17) of the supercharges cannot be generated via a rotation (2.53). Therefore, this is an additional symmetry which occurs only in the Witten model and cannot be found for other $N = 2$ SUSY models in general. Within the Witten model this invariance can be used to fix the overall sign of the SUSY potential by some convention. The standard convention is to choose the overall sign of Φ such that for good SUSY the unique ground state belongs to $\mathcal{H}^-$. That is, the ground state is a negative state (bosonic vacuum). If SUSY is broken the overall sign for Φ may be chosen arbitrarily. As a consequence of this convention we have the following spectral properties of the partner Hamiltonians:

$$\mathrm{spec}(H_-)\backslash\{0\} = \mathrm{spec}(H_+) \qquad \text{for good SUSY,}$$
$$\mathrm{spec}(H_-) = \mathrm{spec}(H_+) \qquad \text{for broken SUSY.}$$

(3.19)

For broken SUSY H_- and H_+ have identical spectrum, whereas for good SUSY H_- and H_+ are only essential iso-spectral. If not explicitly mentioned otherwise we will work within the convention (3.19) from now on.

3.3.3 Asymptotic Behavior of the SUSY Potential

All properties of the Witten model are exclusively determined by the SUSY potential Φ, respectively its integral, the superpotential U. In this section we will show how the asymptotic behavior of Φ and U, respectively, provide an answer to the question whether SUSY is good or broken.

The requirement for a good SUSY is the existence of a normalizable zero-energy wave function $\phi_0^-(x)$ (because of our convention). Thus we have to impose the following condition on the superpotential U:

$$\int_{\mathcal{M}} dx \, \exp\{-2U(x)\} < \infty.$$

(3.20)

In other words, the superpotential has to diverge fast enough for $x \to \pm\infty$,

$$U(x) \to +\infty \qquad \text{for} \qquad x \to \pm\infty.$$

(3.21)

Note that we will discuss only the case $\mathcal{M} = \mathbb{R}$. The discussion for the other cases $\mathcal{M} = \mathbb{R}^+$ and $\mathcal{M} = [a, b]$ are similar.

Let us explore what could be meant by fast enough. For this, we make the ansatz

$$U(x) \sim a_\pm |x|^{\alpha_\pm} \qquad \text{for} \qquad x \to \pm\infty.$$

(3.22)

Clearly, for $a_\pm > 0$ and $\alpha_\pm > 0$ the integral (3.20) exists. Hence, SUSY is good for any superpotential which diverges algebraically to $+\infty$ as $x \to \pm\infty$. The behavior (3.22) implies for the SUSY potential the asymptotic form

$$\Phi(x) = \frac{\hbar}{\sqrt{2m}} U'(x) \sim \frac{\hbar a_\pm \alpha_\pm}{\sqrt{2m}} \frac{|x|^{\alpha_\pm}}{x} \qquad \text{for} \qquad x \to \pm\infty.$$

(3.23)

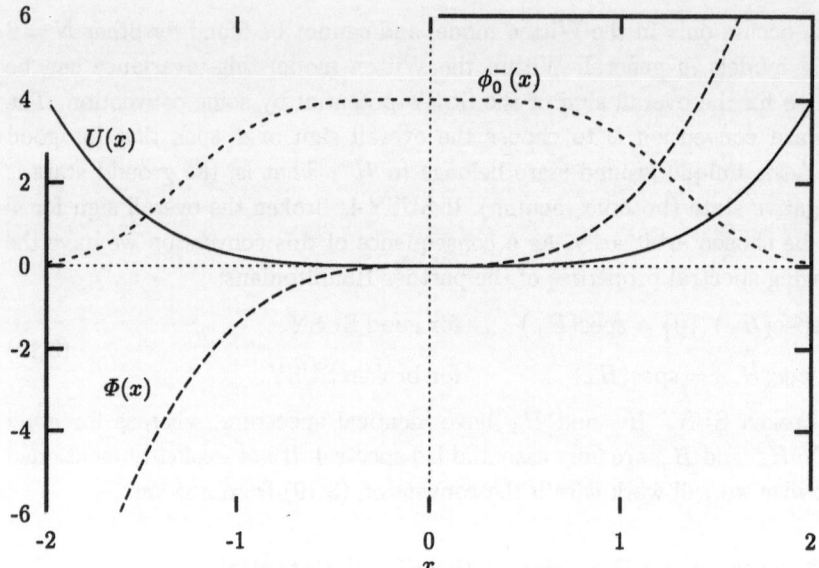

Fig. 3.1. A typical example for good SUSY. Shown are the superpotential $U(x) = x^4/4$, the SUSY potential $\Phi(x)$, and the corresponding normalizable ground-state wave function $\phi_0^-(x) = Ce^{-U(x)}$ in units $\hbar = m = C/4 = 1$.

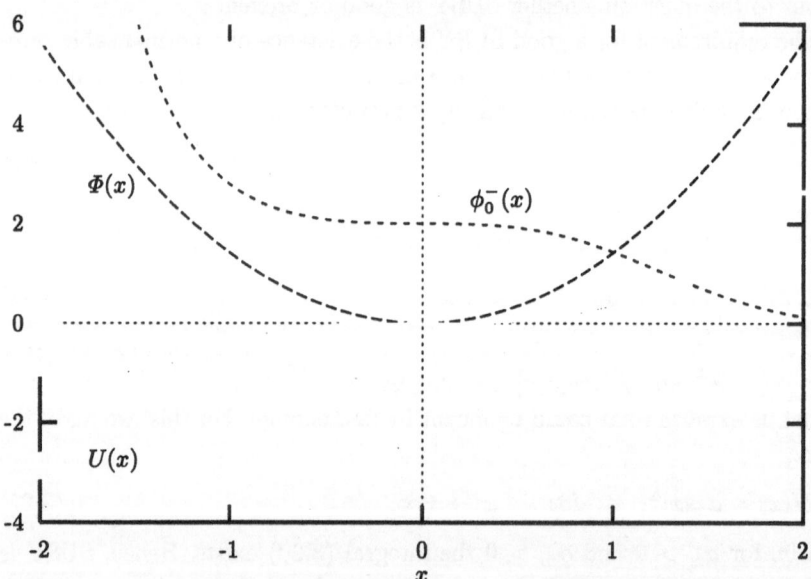

Fig. 3.2. An example for broken SUSY. Again the superpotential $U(x) = x^3/3$, the SUSY potential $\Phi(x)$, and the corresponding non-normalizable function $\phi_0^-(x) = Ce^{-U(x)}$ are shown ($\hbar = m = C/2 = 1$).

In particular, a sufficient condition for good SUSY is $\Phi(x) \to \pm\infty$ as $x \to \pm\infty$ (i.e. $\alpha_\pm > 1$). See Fig. 3.1 for the example $U(x) = x^4/4$. On the contrary, a SUSY potential which for $x \to \pm\infty$ behaves, for example, like $\Phi(x) \to +\infty$ necessarily leads to a broken SUSY. See Fig. 3.2 for the example $U(x) = x^3/3$.

However, even a logarithmic divergence of U at infinity can lead to a good SUSY. Making the ansatz

$$U(x) \sim b_\pm \ln|x| \qquad \text{for} \qquad x \to \pm\infty, \tag{3.24}$$

the square integrability of $\phi_0^-(x)$ requires $b_\pm > \frac{1}{2}$. For the SUSY potential this leads to

$$\Phi(x) \sim \frac{\hbar b_\pm}{\sqrt{2m}} \frac{1}{x} \qquad \text{for} \qquad x \to \pm\infty, \tag{3.25}$$

which may be considered as a limiting case of (3.23) where $\alpha_\pm \to 0$, $a_\pm \to \infty$ such that $b_\pm = \alpha_\pm a_\pm > \frac{1}{2}$.

Let us mention that the conditions $a_\pm > 0$ and $b_\pm > \frac{1}{2}$ are a direct consequence of our convention. If we want to have a positive-parity zero-energy state these conditions change to $a_\pm < 0$ and $b_\pm < -\frac{1}{2}$. In Fig. 3.3 we display the regimes of the parameters $a_\pm$ and $\alpha_\pm$ which give rise to broken and good SUSY, respectively.

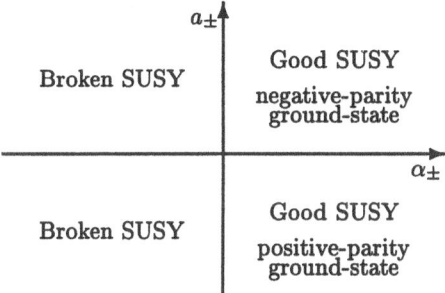

Fig. 3.3. Whether SUSY is broken or not is determined by the asymptotic behavior of the superpotential.

3.4 Broken Versus Good SUSY

In the above discussion we have demonstrated that the breaking of SUSY is completely determined by the asymptotic behavior of the SUSY potential Φ. In fact, more generally, the existence of a zero-energy state is obtainable from the quantities

$$\Phi_\pm := \lim_{x \to \pm\infty} \Phi(x) \in [-\infty, +\infty]. \tag{3.26}$$

Defining the sign function

$$\operatorname{sgn}(z) := \begin{cases} +1 & \text{for} & 0 < z \le +\infty \\ -1 & \text{for} & 0 > z \ge -\infty \end{cases} \qquad (3.27)$$

it is obvious that for $\Phi_\pm \ne 0^1$ we have the equivalence relations

$$\begin{aligned} \operatorname{sgn}(\Phi_+) = \operatorname{sgn}(\Phi_-) &\iff \text{SUSY is broken,} \\ \operatorname{sgn}(\Phi_+) = -\operatorname{sgn}(\Phi_-) &\iff \text{SUSY is good.} \end{aligned} \qquad (3.28)$$

The asymptotic values $\Phi_\pm$ also allow for an explicit representation of the Witten index:

$$\Delta = \tfrac{1}{2}[\operatorname{sgn}(\Phi_+) - \operatorname{sgn}(\Phi_-)]. \qquad (3.29)$$

This result shows that the Witten index does not depend on the details of the SUSY potential Φ. As long as the asymptotic behavior of Φ does not change the sign, that is, $\Phi_\pm$ do not change sign, the Witten index is invariant under deformations of the SUSY potential. This is an example of the "topological invariance" of Δ mentioned in Sect. 2.2.4.

It should be noted that in the case where one or both of the values $\Phi_\pm$ vanish the above criterion (3.28) does not apply. Here more information about the asymptotic behavior of Φ is necessary. See, for example, the discussion of (3.25).

The asymptotic values $\Phi_\pm$ allow also for a discussion of the excited energy eigenvalues of H. For example, if both $|\Phi_+|$ and $|\Phi_-|$ are equal to $+\infty$ then the two SUSY partner Hamiltonians H_+ and H_- have a purely discrete spectrum. On the contrary, if one or both of $|\Phi_\pm|$ are finite (including zero) then the SUSY partner Hamiltonians have, in addition to a possible discrete spectrum, identical continuous spectra starting at $\varepsilon_c := \min(\Phi_+^2, \Phi_-^2)$. In particular, for $\varepsilon_c = 0$ both Hamiltonians $H_\pm$ have a continuous spectrum given by the half line $\mathbb{R}^+$. In addition, if SUSY is good, one of them also has a bound state with eigenvalue $E_0 = \varepsilon_c = 0$.

There exist also other criteria for good and broken SUSY, which of course are equivalent to the above. For example, it has already been noted by Witten [Wit81], that for a continuous differentiable SUSY potential with $|\Phi_\pm| > 0$ SUSY will be broken if Φ has an even number of zeros (counted with their multiplicity). On the other hand, SUSY will be good if Φ has an odd number of zeros. In particular, if Φ is a polynomial of degree p, this number determines whether SUSY is broken (p even) or not (p odd) [JaLeLe87].

Another interesting criterion for broken and good SUSY can be found if the SUSY potential Φ has a definite space parity [InJu93a]. That is, it is an eigenfunction of the space-parity operator Π,

---

[1] This corresponds to $\alpha_\pm \ge 1$ if Φ has an asymptotic behavior as in (3.23).

$$\Pi\Phi(x) := \Phi(-x) = \pm\Phi(x), \tag{3.30}$$

which should not be confused with the Witten-parity operator. Obviously, the operator Π commutes with the tree Hamiltonian H_{tree} defined in (3.7)

$$[\Pi, H_{\text{tree}}] = 0. \tag{3.31}$$

Now, if Φ has even parity, which implies a broken SUSY, then the loop correction H_{loop}, which is proportional to the derivative of Φ, has odd parity and, hence, anticommutes with Π,

$$[\Pi, \Phi] = 0 \quad \Longrightarrow \quad \{\Pi, H_{\text{loop}}\} = 0. \tag{3.32}$$

As a consequence, both Hamiltonians $H_{\pm} = H_{\text{tree}} \pm H_{\text{loop}}$ do not have a well-defined parity. For an even Φ parity as well as SUSY are broken.

If, however, Φ has odd parity then

$$\{\Pi, \Phi\} = 0 \quad \Longrightarrow \quad [\Pi, H_{\text{loop}}] = 0 \quad \Longrightarrow \quad [\Pi, H_{\pm}] = 0 \tag{3.33}$$

and parity as well as SUSY are good symmetries.

3.5 Examples

3.5.1 Systems on the Euclidean Line

As a first class of examples for the Witten model we consider the SUSY quantum mechanics of a cartesian degree of freedom on the Euclidean line $\mathcal{M} = \mathbb{R}$. The corresponding Hilbert subspaces $\mathcal{H}^{\pm} := L^2(\mathbb{R})$ are identical.

Example 1. Supersymmetric anharmonic oscillator:
This system is characterized by the SUSY potential

$$\Phi(x) := \frac{a\hbar}{\sqrt{2m}} \, \text{sgn}(x)|x|^{\alpha-1}, \qquad \alpha > 1, \qquad a > 0, \tag{3.34}$$

which is an odd function of its argument. Hence, by the reasoning of the previous section we expect SUSY to be good. It may easily be verified that the partner potentials are given by

$$V_{\pm}(x) = \frac{\hbar^2}{2m} \left[a^2|x|^{2\alpha-2} \pm a(\alpha-1)|x|^{\alpha-2} \right] = V_{\pm}(-x) \tag{3.35}$$

and the ground-state wave function reads

$$\phi_0^{-}(x) = C \exp\left\{ -\frac{a}{\alpha}|x|^{\alpha} \right\}. \tag{3.36}$$

In Fig. 3.4 we plot for various values of the parameter α the potential V_{-} and V_{+}. Note that $\alpha = 2$ is the special case of the supersymmetric harmonic oscillator and that for large α the potential V_{-} acquires narrow and deep dips

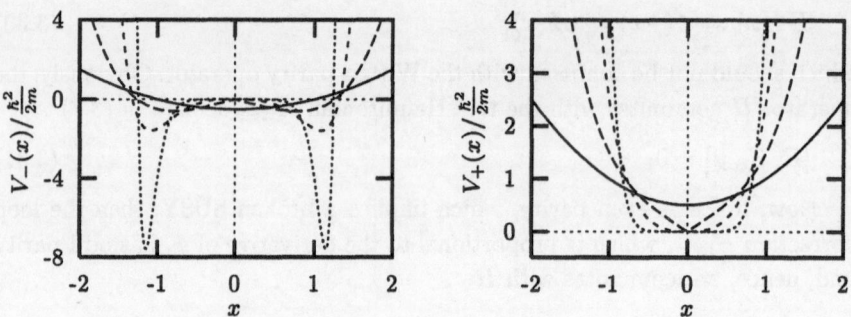

Fig. 3.4. The two partner potentials V_- and V_+ for the good SUSY potential (3.34) for $a = 1$. Displayed are the potentials for $\alpha = 2$ (——), $\alpha = 3$ (– – –), $\alpha = 5$ (- - - -) and $\alpha = 10$ (·····).

near $x = \pm 1$. In fact, in the limit $\alpha \to \infty$ these dips become singular and lead to stationary wave functions with vanishing slope at $x = \pm 1$. In other words, they simulate Neumann boundary conditions, $\phi'(\pm 1) = 0$. For the partner potential V_+ the limit $\alpha \to \infty$ produces Dirichlet conditions at $x = \pm 1$, that is, $\phi(\pm 1) = 0$.

Example 2. Anharmonic oscillator with broken SUSY:
Dropping the sign function in (3.34) we obtain the SUSY potential

$$\Phi(x) := \frac{a\hbar}{\sqrt{2m}}\, |x|^{\alpha-1}, \qquad a > 0, \qquad \alpha > 1, \tag{3.37}$$

which now is an even function giving rise to a broken SUSY. The corresponding partner potentials read

$$V_\pm(x) = \frac{\hbar^2}{2m} \left[a^2 |x|^{2\alpha-2} \pm a(\alpha-1)\mathrm{sgn}(x)|x|^{\alpha-2} \right]. \tag{3.38}$$

Note that $V_\pm(-x) = V_\mp(x)$ and, hence, it is not surprising that the two partner Hamiltonians have identical spectrum as expected from broken SUSY. We have plotted V_- in Fig. 3.5 for the same parameters as in the previous example. Note that besides SUSY also parity is broken. Here in the limit $\alpha \to \infty$ the potential V_- simulates at $x = -1$ a Dirichlet condition and at $x = +1$ a Neumann condition (and vice versa for V_+).

Example 3. The free particle and its SUSY partner:
As a third example we consider the SUSY potential

$$\Phi(x) := \frac{\hbar}{\sqrt{2m}}\, \tanh(x), \tag{3.39}$$

which gives rise to the two partner potentials

$$V_+(x) = \frac{\hbar^2}{2m}, \qquad V_-(x) = \frac{\hbar^2}{2m} \left[1 - \frac{2}{\cosh^2 x} \right]. \tag{3.40}$$

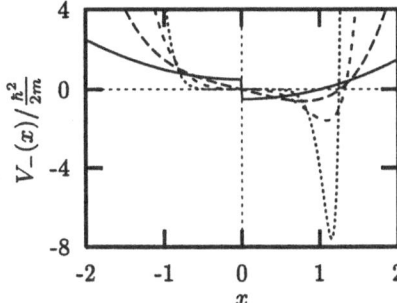

Fig. 3.5. The same as Fig. 3.4 but for the broken-SUSY potential (3.37). Here only V_- is plotted as $V_+(x) = V_-(-x)$.

Again SUSY is good as $\Phi_+ = -\Phi_-$ (cf. eq. (3.28)). The potential V_+ is constant and hence characterizes a free particle. There are no bound states for H_+ and therefore, because of good SUSY, H_- has a single bound state with zero energy. The corresponding wave function reads

$$\phi_0^-(x) = \frac{C}{\cosh(x)}. \tag{3.41}$$

The potential V_- is an example of the so-called symmetric Rosen-Morse potentials. The Schrödinger equation for this type of potentials is studied, for example, in connection with soliton solutions of the Korteweg-de Vries equation [Tha92] or relaxation processes in Fermi liquids [VoVoTo85]. It is easily shown that the potential V_- is reflectionless due to the reflectionlessness of its SUSY partner V_+, the constant potential [CoKhSu95]. In Fig. 3.6 we have plotted both potentials.

Example 4. Attractive and repulsive δ-potential:
As a last example on the Euclidean line we mention the attractive and repulsive δ-potential. For this we consider the SUSY potential of the form

$$\Phi(x) := \frac{a\hbar}{\sqrt{2m}} \left[\Theta(x) - \Theta(-x) \right], \qquad a > 0. \tag{3.42}$$

Note that this SUSY potential is not continuously differentiable, an assumption we have made so far for all SUSY potentials. Nevertheless, interpreting the derivative of the unit-step function in a distributional sense, that is, $\Theta'(x) = \delta(x)$, we arrive at

$$V_\pm(x) = \frac{\hbar^2}{2m} \left[a^2 \pm 2a\delta(x) \right], \qquad \phi_0^-(x) = C \exp\{-a|x|\}. \tag{3.43}$$

The attractive and repulsive δ-potential are SUSY partners. Note that SUSY is good as the bound state associated with V_- has zero energy. This model can also be considered as the limiting case $\alpha = 1$ of our first example.

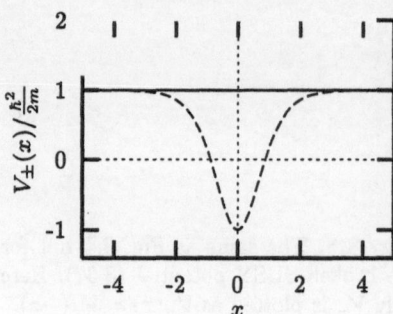

Fig. 3.6. The free particle on the real line (——) and its SUSY partner (- - -).

3.5.2 Systems on the Half Line

The second class of examples, which we are going to mention, has as configuration space the positive half line $\mathcal{M} = \mathbb{R}^+$, that is, $x \geq 0$. Here the partner Hamiltonians $H^\pm$ are formal differential operators and require a careful specification of their domains $\mathcal{H}^\pm$, which in fact depend on the explicit form of the potentials $V_\pm$. In the following we will consider the Hilbert space with Dirichlet boundary condition at $x = 0$,

$$\mathcal{H}^\pm := \{\phi \in L^2(\mathbb{R}^+) \ : \ \phi(0) = 0\}, \tag{3.44}$$

and only those potentials $V_\pm$ for which $H_\pm$ will be self-adjoint on $\mathcal{H}^\pm$. We note that on $\mathcal{H}^\pm$ the momentum operator p is symmetric but not self-adjoint [Ric78]. Hence, the supercharges Q_1 and Q_2 are also symmetric but in general not self-adjoint. However, $p^\dagger p$ is still self-adjoint on $\mathcal{H}^\pm$ and therefore, also $H_\pm$ may be well-defined partner Hamiltonians if the SUSY potential Φ is chosen appropriately. For example, a singular SUSY potential which behaves like $\Phi(x) \sim -\eta/x$ near $x = 0$ requires a careful discussion [JeRo84, Fuc86, ShSmVa88] showing that not all $\eta \in \mathbb{R}$ are admissible if the Dirichlet condition in (3.44) would have been omitted. If not such care is taken, then one may find negative energy eigenstates or unpaired positive-energy eigenvalues for $H_\pm$ [Cas91, PaSu93] which are a direct consequence of the non-selfadjointness of the supercharges.

Below, for each example, we simply list the SUSY potential, the partner potentials, the ground-state wave function, and then make some comments.

Example 1. Radial harmonic oscillator:

$$\Phi(x) := \sqrt{\frac{m}{2}}\,\omega x - \frac{\hbar}{\sqrt{2m}}\,\frac{\eta}{x}, \qquad \omega > 0,$$

$$V_\pm(x) = \frac{m}{2}\,\omega^2 x^2 + \frac{\hbar^2 \eta(\eta \pm 1)}{2mx^2} - \hbar\omega\left(\eta \pm \frac{1}{2}\right), \qquad (3.45)$$

$$\phi_0^-(x) = Cx^\eta \exp\left\{-\frac{m\omega}{2\hbar}\,x^2\right\}.$$

Obviously, for $\eta > 0$ SUSY is good. In particular, if we set $\eta = l + 1$, $l \in \mathbb{N}_0$, the two potentials $V_\pm$ are identical to those of the radial harmonic oscillator for fixed angular momentum l and $l + 1$, respectively. On the other hand, for $\eta \leq 0$ we have a broken SUSY. Despite the fact, that $\eta \in (-1/2, 0]$ does lead to a square integrable ground-state wave function, it does not obey the boundary condition $\phi_0^-(0) = 0$. Because of this boundary condition the partner Hamiltonian associated with the above potentials $V_\pm$ are indeed self-adjoint. In fact, we have chosen a particular self-adjoint extension of the formal operators $H_\pm$ by (3.44). This is true for any $\eta \in \mathbb{R}$ because $\eta(\eta \pm 1) \geq -1/4$ [ReSi75]. In the broken case $\eta \leq 0$ we point out that for $\eta = -l$ we again recover (up to some constant) the same two radial harmonic oscillators with fixed angular momentum l and $l - 1$. Hence, the radial harmonic oscillator is a supersymmetric quantum system with good or broken SUSY, depending on the choice of the parameter η.

Example 2. The modified Pöschl–Teller problem:

$$\Phi(x) := \frac{\hbar}{\sqrt{2m}}\,[\kappa \tanh x - \eta \coth x], \qquad \kappa > 0,$$

$$V_\pm(x) = \frac{\hbar^2}{2m}\left[\frac{\eta(\eta \pm 1)}{\sinh^2 x} - \frac{\kappa(\kappa \pm 1)}{\cosh^2 x} + (\kappa - \eta)^2\right], \qquad (3.46)$$

$$\phi_0^-(x) = C\,\frac{\sinh^\kappa x}{\cosh^\eta x}.$$

The above potentials $V_\pm$ are called modified Pöschl–Teller potentials. Here SUSY is good if $0 < \eta < \kappa$. As in the above example the partner potentials are up to an additive constant invariant under the replacement $\eta \to -\eta$. What does change is the nature of SUSY, that is, good SUSY becomes broken and vice versa.

Example 3. Radial hydrogen atom:

$$\Phi(x) := \sqrt{\frac{2m}{\hbar}} \frac{\alpha}{2\eta} - \frac{\hbar}{\sqrt{2m}} \frac{\eta}{x}, \qquad \alpha, \eta > 0,$$

$$V_\pm(x) = -\frac{\alpha}{x} + \frac{\hbar^2 \eta(\eta \pm 1)}{2mx^2} + \frac{m\alpha^2}{2\eta^2 \hbar^2}, \tag{3.47}$$

$$\phi_0^-(x) = Cx^\eta \exp\left\{-\frac{m\alpha}{2\hbar^2} x\right\}.$$

Note that because of our convention (3.19) we have to impose the condition $\eta > 0$. For $\eta = l + 1$ we recognize the radial hydrogen problem. Supersymmetric aspects of the one-dimensional hydrogen atom are discussed in [SiTePoLu90].[2]

3.5.3 Systems on a Finite Interval

As a last class of examples for the generalized Witten model we consider those which are defined on a finite interval $x \in \mathcal{M} = [a, b]$. Here, in contrast to the previous class, the momentum operator and hence the supercharges can be made self-adjoint if appropriate boundary conditions are specified at $x = a$ and b.

Example 1. Infinite square well and its SUSY partner:
The first example we mention is the infinite square well of width π chosen for convenience symmetric about the origin:

$$\Phi(x) := \frac{\hbar}{\sqrt{2m}} \tan x, \qquad x \in [-\pi/2, \pi/2],$$

$$V_+(x) = \frac{\hbar^2}{2m} \left(\frac{2}{\cos^2 x} - 1\right), \qquad V_-(x) = -\frac{\hbar^2}{2m}, \tag{3.48}$$

$$\phi_0^-(x) = C \cos x,$$

with domain $\mathcal{H}^\pm := \{\phi \in L^2([-\pi/2, \pi/2]) : \phi(-\pi/2) = 0 = \phi(\pi/2)\}$. Hence, the SUSY partner for the particle in a box is given by the symmetric Pöschl–Teller potential V_+. SUSY in this case is a good symmetry. Note that the square well V_- is shifted such that $E_0 = 0$.

---

[2] There has been a long-lasting discussion about the spectral properties of the one-dimensional hydrogen atom in the literature. For a clarification see [FiLeMü95].

Example 2. The non-symmetric Pöschl-Teller oscillator:

$$\Phi(x) := \frac{\hbar}{\sqrt{2m}} (\kappa \tan x - \eta \cot x), \qquad \kappa > 0, \qquad x \in [0, \pi/2],$$

$$V_\pm(x) = \frac{\hbar^2}{2m} \left[\frac{\kappa(\kappa \pm 1)}{\cos^2 x} + \frac{\eta(\eta \pm 1)}{\sin^2 x} - (\kappa + \eta)^2 \right], \tag{3.49}$$

$$\phi_0^-(x) = C \cos^\kappa x \sin^\eta x,$$

with domain $\mathcal{H}^\pm := \{\phi \in L^2([0, \pi/2]) : \phi(0) = 0 = \phi(\pi/2)\}$. Obviously, the non-symmetric Pöschl-Teller oscillators with parameter set (κ, η) and $(\kappa - 1, \eta - 1)$ are SUSY partners. SUSY is good for $\eta > 0$ and broken for $\eta \le 0$.

Example 3. The supersymmetric square well:
Here we assume a vanishing SUSY potential in the configuration space $\mathcal{M} := [-1, 1]$ and consequently also the partner potentials vanish on $\mathcal{M}$

$$\Phi(x) := 0, \qquad V_\pm(x) = 0. \tag{3.50}$$

The SUSY structure is introduced via the boundary conditions at $x = \pm 1$:

$$\mathcal{H}^+ := \{\phi \in L^2([-1, 1]) : \phi(-1) = 0 = \phi(1)\},$$
$$\mathcal{H}^- := \{\phi \in L^2([-1, 1]) : \phi'(-1) = 0 = \phi'(1)\}. \tag{3.51}$$

That is, for $H_+ = p^2/2m$ we choose Dirichlet conditions at $x = \pm 1$. Whereas, for $H_- = p^2/2m$ we take Neumann conditions at those points. These imply a good SUSY as the ground-state wave function is constant,

$$\phi_0^-(x) = 1/\sqrt{2}. \tag{3.52}$$

This example can be considered as the limiting case $\alpha \to \infty$ of (3.34), the supersymmetric anharmonic oscillator.

Example 4. Square well with broken SUSY:
As in the previous example we assume a vanishing SUSY potential on $\mathcal{M} = [-1, 1]$. However, the breaking of SUSY is introduced via non-symmetric boundary conditions at $x = \pm 1$:

$$\mathcal{H}^+ := \{\phi \in L^2([-1, 1]) : \phi'(-1) = 0 = \phi(1)\},$$
$$\mathcal{H}^- := \{\phi \in L^2([-1, 1]) : \phi(-1) = 0 = \phi'(1)\}. \tag{3.53}$$

That is, for $H_+ = p^2/2m$ we take a Neumann condition at $x = -1$ and a Dirichlet condition at $x = 1$, and vice versa for H_-. Clearly, parity as well as SUSY are broken. This example may be understood as the $\alpha \to \infty$ limit of (3.37).

4. Supersymmetric Classical Mechanics

In the present chapter we consider the classical version of the supersymmetric Witten model. Actually, these kinds of supersymmetric classical systems can be considered as special cases of so-called *pseudoclassical* models. The notion of pseudoclassical mechanics has been introduced in 1976 by Casalbuoni [Cas76b] and describes classical systems which in addition to their usual bosonic (commuting) also have fermionic (anticommuting) degrees of freedom. In fact, pseudoclassical systems can be understood as the classical limit of quantum systems which have both kinds of degree of freedom [Cas76b]. Pseudoclassical mechanics is of particular interest because of its ability to describe spin-degrees of freedom on a classical level [BeMa75, Cas76a, BaCaLu76, BrDeZuVeHo76, BeMa77, BaBoCa81, LaDoGu93]. For an alternative classical spin model see [BaZa84, Bar86].

4.1 Pseudoclassical Models

The construction of pseudoclassical models has systematically been discussed by Casalbuoni [Cas76c]. Here we will study the properties of the simplest non-trivial example which allows for an interaction between bosonic and fermionic degrees of freedom. It is characterized by the pseudoclassical Lagrangian ($m = 1$):

$$L_0 := \tfrac{1}{2}\dot{x}^2 - V_1(x) + \tfrac{i}{2}(\bar{\psi}\dot{\psi} - \dot{\bar{\psi}}\psi) - V_2(x)\bar{\psi}\psi. \tag{4.1}$$

In the above x denotes a bosonic degree of freedom and ψ and $\bar{\psi}$ denote independent fermionic degrees of freedom. The potentials V_1 and V_2 are differentiable functions of x.

Let us first discuss the nature of the bosonic and fermionic degrees of freedom. They are not represented by ordinary c-numbers but by even and odd elements of a Grassmann algebra. In the present case this algebra may be constructed via two generators denoted by ψ_0 and $\bar{\psi}_0$ which obey the following relations[1]

$$\{\psi_0, \bar{\psi}_0\} = 0, \qquad \psi_0^2 = 0 = \bar{\psi}_0^2. \tag{4.2}$$

---

[1] For details about Grassmann numbers and Grassmann algebras see, for example, [Ber66, DeWit92, Cor89, CoGr94].

The overbar denotes the adjoint Grassmann number. For an arbitrary element of the Grassmann algebra

$$B := a_1 + a_2\psi_0 + a_3\bar{\psi}_0 + a_4\bar{\psi}_0\psi_0, \qquad a_i \in \mathbb{C}, \tag{4.3}$$

its adjoint is defined as

$$\bar{B} := a_1^* + a_2^*\bar{\psi}_0 + a_3^*\psi_0 + a_4^*\bar{\psi}_0\psi_0. \tag{4.4}$$

Here, the mapping $\psi \mapsto \bar{\psi}$ is linear and involutive [Ber66]. Even elements of this algebra are those which commute with all Grassmann numbers of the form (4.3). The bosonic degree of freedom x is, therefore, expected to be of the form

$$x = x_1 + x_2\bar{\psi}_0\psi_0, \tag{4.5}$$

where $x_1, x_2 \in \mathbb{R}$ which assures the reality of the bosonic variable $\bar{x} = x$. As x is a real and even Grassmann number, as given in (4.5), so are the potentials V_i:

$$V_i(x) = V_i(x_1) + V_i'(x_1)x_2\bar{\psi}_0\psi_0, \qquad i = 1, 2. \tag{4.6}$$

On the other hand, the fermionic degrees of freedom are odd elements of the Grassmann algebra

$$\psi = a\psi_0 + b\bar{\psi}_0, \qquad \bar{\psi} = a^*\bar{\psi}_0 + b^*\psi_0 \tag{4.7}$$

with $a, b \in \mathbb{C}$.

4.2 A Supersymmetric Classical Model

Now we will specialize the above Lagrangian (4.1) by setting

$$V_1(x) := \tfrac{1}{2}\Phi^2(x), \qquad V_2(x) := \Phi'(x), \tag{4.8}$$

where, as in the sense above,

$$\Phi(x) = \Phi(x_1) + \Phi'(x_1)x_2\bar{\psi}_0\psi_0. \tag{4.9}$$

It will be shown later that Φ is up to a constant factor $\sqrt{2}$ identical to the SUSY potential in Witten's model.

The special class of systems we are now dealing with is characterized by a Lagrangian of the form

$$L := \tfrac{1}{2}\dot{x}^2 - \tfrac{1}{2}\Phi^2(x) + \tfrac{i}{2}\left(\bar{\psi}\dot{\psi} - \dot{\bar{\psi}}\psi\right) - \Phi'(x)\bar{\psi}\psi. \tag{4.10}$$

This Lagrangian characterizes pseudoclassical systems being supersymmetric. Indeed, this model can be derived via a supersymmetrization of a $(0+1)$-dimensional field theory. By supersymmetrization we mean the extension of the $(0+1)$-dimensional space spanned by the time variable t to a superspace[2] spanned by $(t, \theta, \bar{\theta})$ where θ and $\bar{\theta}$ are odd Grassmann variables. The general procedure has been outlined by Nicolai [Nic76] and is basically a special case of the supersymmetrization of $(3+1)$-dimensional field theories [WeZu74a, WeZu74b]. For a detailed discussion based on the work of Nicolai [Nic76] see, for example, [CoFr83] or Sect. 2 in [Mis91]. In essence, one first arrives at a Lagrangian density $\mathcal{L}$ over the superspace $(t, \theta, \bar{\theta})$ being invariant under SUSY transformations, which consist basically of translations generated by $\partial/\partial t$, $\partial/\partial\theta$ and $\partial/\partial\bar{\theta}$. The "effective" Lagrangian (4.10) is then obtained via integration, $L = \int d\theta d\bar{\theta}\, \mathcal{L}$, and characterizes a system being invariant under the following SUSY transformation of the fields [Nic76]:

$$x \mapsto x + \delta x, \qquad \delta x := \bar{\varepsilon}\psi + \bar{\psi}\varepsilon,$$

$$\psi \mapsto \psi + \delta\psi, \qquad \delta\psi := -(\mathrm{i}\dot{x} + \Phi(x))\varepsilon, \qquad (4.11)$$

$$\bar{\psi} \mapsto \bar{\psi} + \delta\bar{\psi}, \qquad \delta\bar{\psi} := (\mathrm{i}\dot{x} - \Phi(x))\bar{\varepsilon},$$

where ε and $\bar{\varepsilon}$ denote "infinitesimal" versions of the Grassmann variables θ and $\bar{\theta}$. The invariance of such systems under the above SUSY transformation becomes obvious by noting that (4.11) implies

$$L \mapsto L + \delta L, \qquad \delta L = \frac{1}{2}\frac{d}{dt}\left[(\dot{x} - \mathrm{i}\Phi)\bar{\varepsilon}\psi + (\dot{x} + \mathrm{i}\Phi)\bar{\psi}\varepsilon\right], \qquad (4.12)$$

which gives rise to a Lagrangian $L + \delta L$ being gauge-equivalent to the original Lagrangian L. In other words, $L + \delta L$ and L characterize the same pseudo-classical system.

4.3 The Classical Dynamics

Despite the fact that these pseudoclassical and supersymmetric classical models are known for almost 20 years, it has only recently been observed [JuMa94, JuMa95] that the solution of the corresponding classical equations of motion can be derived in a rather explicit form and have some interesting properties.

Let us start with the equations of motion derived from (4.10):

$$\dot{\psi} = -\mathrm{i}\Phi'(x)\psi, \qquad \dot{\bar{\psi}} = \mathrm{i}\Phi'(x)\bar{\psi}, \qquad (4.13)$$

$$\ddot{x} = -\Phi(x)\Phi'(x) - \Phi''(x)\bar{\psi}\psi. \qquad (4.14)$$

---

[2] The superspace formulation has been introduced by Salam and Strathdee [SaSt74]. For the properties of superspaces see, for example, [DeWit92, Cor89, CoGr94].

The first-order differential equations for the fermionic degrees of freedom can immediately be integrated. With initial conditions $\psi_0 := \psi(0)$ and $\bar{\psi}_0 := \bar{\psi}(0)$ integration of (4.13) gives

$$\psi(t) = \psi_0 \exp\left\{-i \int_0^t d\tau\, \Phi'(x(\tau))\right\},$$

$$\bar{\psi}(t) = \bar{\psi}_0 \exp\left\{i \int_0^t d\tau\, \Phi'(x(\tau))\right\}, \qquad (4.15)$$

where x denotes the (yet unknown) solution of (4.14). Let us note that the solutions (4.15) imply that $\bar{\psi}(t)\psi(t) = \bar{\psi}_0\psi_0$ is a constant and, therefore, eq. (4.14) simplifies to

$$\ddot{x} = -\Phi(x)\Phi'(x) - \Phi''(x)\bar{\psi}_0\psi_0. \qquad (4.16)$$

As the SUSY potential Φ is assumed to be of the form (4.9), the bosonic degree of freedom x is necessarily of the form (4.5):

$$x(t) = x_{qc}(t) + q(t)\bar{\psi}_0\psi_0, \qquad (4.17)$$

where x_{qc} and q are real-valued functions of time. We will call x_{qc} the *quasi-classical* solution in order to differentiate it from the full pseudoclassical solution x which contains the $\bar{\psi}_0\psi_0$–term and in general is an even-Grassmann-valued function of time. It is also worth mentioning that in (4.15) one may replace $x(\tau)$ by $x_{qc}(\tau)$ because of (4.17) and (4.2). Hence, we have

$$\psi(t) = \psi_0 \exp\left\{-2i\varphi[x_{qc}]\right\}, \qquad \bar{\psi}(t) = \bar{\psi}_0 \exp\left\{2i\varphi[x_{qc}]\right\}, \qquad (4.18)$$

where we have introduced the *fermionic phase*

$$\boxed{\varphi[x] := \tfrac{1}{2} \int_0^t d\tau\, \Phi'\Big(x(\tau)\Big),} \qquad (4.19)$$

a functional which we will revisit in our quasi-classical approximation to the quantum propagator for the Witten model.

Now we study the solution for the bosonic degree of freedom. Multiplication of (4.16) with $\dot{x}$ and integration leads to the conservation of the energy

$$\mathcal{E} = \tfrac{1}{2}\dot{x}^2 + \tfrac{1}{2}\Phi^2(x) + \Phi'(x)\bar{\psi}_0\psi_0, \qquad (4.20)$$

which is a constant even Grassmann number. The ansatz (4.17) together with $\mathcal{E} =: E + F\bar{\psi}_0\psi_0$ ($E \geq 0$ and $F \in \mathbb{R}$ are constants) results in

$$\dot{x}_{qc}^2 = 2E - \Phi^2(x_{qc}), \qquad (4.21)$$

$$\dot{q} = \frac{1}{\dot{x}_{qc}}\left[F - \Phi'(x_{qc}) - \Phi(x_{qc})\Phi'(x_{qc})q\right]. \qquad (4.22)$$

The last equation, which determines $q(t)$ if $\dot{x}_{qc} \not\equiv 0$, can also be integrated and yields [JuMa94]

$$q(t) = \frac{\dot{x}_{qc}(t)}{\dot{x}_{qc}(0)} q(0) + \dot{x}_{qc}(t) \int\limits_0^t d\tau\, \frac{F - \Phi'(x_{qc}(\tau))}{2E - \Phi^2(x_{qc}(\tau))}, \tag{4.23}$$

where $q(0)$ is a constant of integration. Again we find, as for the fermionic degrees of freedom, that $q(t)$ is expressible in terms of the quasi-classical solution $x_{qc}(t)$ determined by (4.21). Let us note that the singularity of the integral in (4.23) near the turning points of the quasi-classical path is precisely canceled by its prefactor as $\dot{x}_{qc}(t)$ vanishes at those points. Hence, $q(t)$ remains finite for all $t \geq 0$. Let us also note that even for the initial condition $q(0) = 0$ we have in general $q(t) \neq 0$ for $t > 0$. In other words, even assuming the pseudoclassical solution initially to be real, $x(0) \in \mathbb{R}$, it will in general become a Grassmann-valued quantity. It is only in the special case $\Phi'(x) = F = $ const., that is, for a linear SUSY potential Φ, where a real $x(0)$ remains to be real for ever.

Let us now discuss some properties of the quasi-classical solution $x_{qc}(t)$. The equation of motion (4.21) for the quasi-classical path can be obtained from a "quasi-classical" Lagrangian defined by

$$L_{qc} := \tfrac{1}{2}\dot{x}^2 - \tfrac{1}{2}\Phi^2(x) = \tfrac{1}{2}\left(\dot{x} \pm i\Phi(x)\right)^2 \mp i\Phi(x)\dot{x}. \tag{4.24}$$

Here x will denote the usual real-valued degree of freedom. The second equality above shows that this Lagrangian is gauge-equivalent to

$$\tilde{L}_{qc}^{\pm} := \tfrac{1}{2}\left(\dot{x} \pm i\Phi(x)\right)^2. \tag{4.25}$$

As an aside we note that the above complex gauge transformations $L_{qc} \to \tilde{L}_{qc}^{\pm}$ become real by using Euclidean instead of real time.

The canonical momenta obtained from the two Lagrangians $\tilde{L}_{qc}^{\pm}$ are

$$\xi^{\pm} := \frac{\partial \tilde{L}_{qc}^{\pm}}{\partial \dot{x}} = \dot{x} \pm i\Phi(x) = \left(\xi^{\mp}\right)^* \tag{4.26}$$

and, surprisingly, coincide with the generators of the SUSY transformation (4.11) of the fermionic degrees of freedom:

$$\delta\psi = -i\xi^-\bar{\varepsilon}, \qquad \delta\bar{\psi} = i\xi^+\varepsilon. \tag{4.27}$$

In fact, with these canonical momenta one can construct classical supercharges [Nic91]

$$Q := \frac{i}{\sqrt{2}}\xi^-\bar{\psi}, \qquad \bar{Q} := -\frac{i}{\sqrt{2}}\xi^+\psi. \tag{4.28}$$

It is easily verified with the help of (4.13) and (4.14) that these supercharges are constants of motion. Indeed, from (4.25) we derive the equations of motion

$$\dot{\xi}^{\pm} = \mp i\Phi'(x_{\mathrm{qc}})\xi^{\pm}, \tag{4.29}$$

which are identical in form with those for the fermionic degrees of freedom. Obviously, the solutions read

$$\xi^{\pm}(t) = \xi^{\pm}(0)\exp\{\mp 2i\varphi[x_{\mathrm{qc}}]\} \tag{4.30}$$

and explicate the conservation of the supercharges (4.28). It is also obvious that the conserved energy E of the quasi-classical solution can be expressed in terms of these canonical momenta, $E = \frac{1}{2}\xi^{+}\xi^{-}$. As a consequence we have the relation

$$\xi^{\pm}/\sqrt{2E} = \left(\xi^{\mp}/\sqrt{2E}\right)^{-1} \tag{4.31}$$

valid for $E > 0$.

For $E = 0$ the quasi-classical solutions are given by $x_{\mathrm{qc}}(t) = x_k$, where x_k are the zeros of the potential Φ, $\Phi(x_k) = 0$. These are precisely the classical ground states for good SUSY. SUSY will be broken on the classical level if the SUSY potential Φ does not have zeros, because then $E \geq \frac{1}{2}\Phi^2(x) > 0$ for all $x \in \mathbb{R}$. See also Fig. 4.2 below. Note that the solution (4.23) for q may not be used in this case. Actually, because of $\dot{x}_{\mathrm{qc}} = \Phi(x_k) = 0$ the equation of motion corresponding to (4.22) reads

$$F = \Phi'(x_{\mathrm{qc}}) = \Phi'(x_k). \tag{4.32}$$

As a consequence, the solutions of the fermionic degrees of freedom are given by

$$\psi(t) = \psi_0\,e^{-itF}, \qquad \bar{\psi}(t) = \bar{\psi}_0\,e^{-itF}. \tag{4.33}$$

There is no equation of motion for q and hence it remains undetermined. This indicates that q can't be a physical degree of freedom.

Finally, let us make two remarks. First we note, that the above procedure for solving the dynamics of the supersymmetric classical system (4.10) can also be used to derive [JuMa95] solutions for the more general pseudoclassical system (4.1).

In a second remark we point out that the gauge transformations $L_{\mathrm{qc}} \to \tilde{L}^{\pm}_{\mathrm{qc}}$ are the classical analogues of the so-called Nicolai map [Nic80a, Nic80b]. This map characterizes a transformation of bosonic fields of a SUSY quantum field theory. It has the interesting property that the full bosonic action is mapped into a free action of boson fields. Note that the quasi-classical Lagrangians $\tilde{L}^{\pm}_{\mathrm{qc}}$ are quadratic in their canonical momenta $\xi^{\pm}$. For an extensive review on the Nicolai map see Ezawa and Klauder [EzKl85] and Sect. 7.3.

Let us now discuss in detail the fermionic phase defined in (4.19). This discussion will be presented in a separate section because of its fundamental role played in the quasi-classical approximation considered in Sect. 6.1.2.

4.4 Discussion of the Fermionic Phase

According to its definition (4.19) the fermionic phase in (4.18) is a functional of the quasi-classical path. For simplicity let us assume that the quasi-classical solution starts with positive velocity $\dot{x}_{qc}(0) > 0$ and t is sufficiently small such that $\dot{x}_{qc}(\tau) > 0$ for all $\tau < t$. In this case the fermionic phase can easily be calculated using the equation of motion (4.21):

$$\varphi[x_{qc}] = \frac{1}{2} \int_{x_{qc}(0)}^{x_{qc}(t)} dx \, \frac{\Phi'(x)}{\sqrt{2E - \Phi^2(x)}} = \frac{1}{2}[a(x'') - a(x')] \qquad (4.34)$$

where, for convenience, we have introduced the abbreviations

$$a(x) := \arcsin\frac{\Phi(x)}{\sqrt{2E}} \in \left[-\frac{\pi}{2}, \frac{\pi}{2}\right],$$
$$x'' := x_{qc}(t), \qquad x' := x_{qc}(0). \qquad (4.35)$$

If we further assume that we initially start at a zero of the SUSY potential, that is, $\Phi(x') = 0$, the fermionic solution ψ takes the simple form

$$\psi(t) = \psi_0 \exp\{-i \arcsin\frac{\Phi(x'')}{\sqrt{2E}}\}$$
$$= \psi_0 \left[\left(1 - \frac{\Phi^2(x'')}{2E}\right)^{1/2} - i\frac{\Phi(x'')}{\sqrt{2E}}\right] \qquad (4.36)$$
$$= \frac{\psi_0}{\sqrt{2E}}[\dot{x}_{qc}(t) - i\Phi(x_{qc}(t))].$$

Hence, the fermionic degrees of freedom are expressible in terms of the canonical momenta $\xi^{\pm}$ of the quasi-classical Lagrangians $\widetilde{L}_{qc}^{\pm}$ as expected from relation (4.30):

$$\psi(t) = \frac{1}{\sqrt{2E}}\xi^{-}(t)\psi_0 = \frac{\psi_0}{2E}\xi^{+}(0)\xi^{-}(t),$$
$$\bar{\psi}(t) = \frac{1}{\sqrt{2E}}\xi^{+}(t)\bar{\psi}_0 = \frac{\bar{\psi}_0}{2E}\xi^{-}(0)\xi^{+}(t). \qquad (4.37)$$

These are the finite versions of the infinitesimal SUSY transformation (4.27). Note that because of the assumption $\Phi(x') = 0$ we have $\xi^{+}(0) = \xi^{-}(0)$ and therefore $\xi^{\pm}(0) = 1/\sqrt{2E}$.

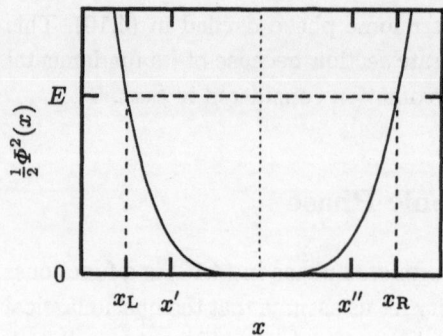

Fig. 4.1. A typical shape of the potential $\frac{1}{2}\Phi^2$. The quasi-classical motion for a given energy E starts at $x_{qc}(0) = x'$ and reaches at time t the position $x_{qc}(t) = x''$. During its motion the left and right turning points x_L and x_R may be visited several times.

Let us now consider a more general quasi-classical solution which also passes turning points. These are points where $\dot{x}_{qc}$ vanishes. The only assumption we make is that Φ^2 has one global minimum and diverges for $x \to \pm\infty$. See Fig. 4.1 for a typical shape of $\frac{1}{2}\Phi^2$. In other words, we assume the quasi-classical solution to be a bounded periodic motion about this minimum. For the SUSY potential this assumption implies that it has at most one zero which, however, may have a multiplicity larger than one. Typical SUSY potentials with good and broken SUSY at the classical as well as quantum level [Wit81] are shown in Fig. 4.2.

Being a one-dimensional system we can rather explicitly discuss the possible quasi-classical paths for a given energy $E > 0$. It is convenient [Schu81] to consider the following four classes. Without loss of generality, we may assume $x'' > x'$, see Fig. 4.1.

(1) Paths which leave x' to the right and reach x'' from the left:
     $\dot{x}_{qc}(0) > 0$, $\dot{x}_{qc}(t) > 0$.

(2) Paths which leave x' to the left and reach x'' from the left:
     $\dot{x}_{qc}(0) < 0$, $\dot{x}_{qc}(t) > 0$.

(3) Paths which leave x' to the right and reach x'' from the right:
     $\dot{x}_{qc}(0) > 0$, $\dot{x}_{qc}(t) < 0$.

(4) Paths which leave x' to the left and reach x'' from the right:
     $\dot{x}_{qc}(0) < 0$, $\dot{x}_{qc}(t) < 0$.

Within each class the paths are uniquely characterized by their number of complete cycles they perform before arriving at x'' at time t. Each of these full cycles contributes a term $a(x_R) - a(x_L)$ to the fermionic phase, where x_R and x_L are the quasi-classical right and left turning points of the periodic motion with energy E:

$$\Phi^2(x_R) = 2E = \Phi^2(x_L). \tag{4.38}$$

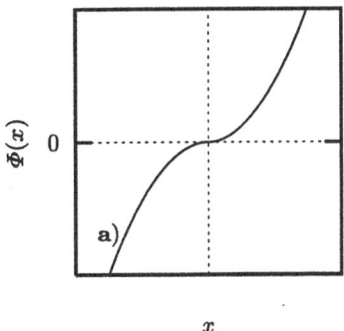

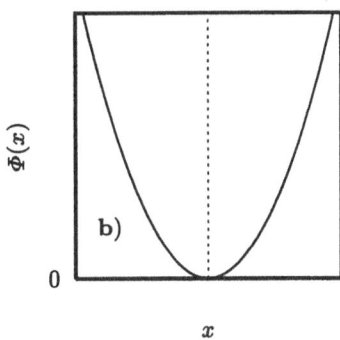

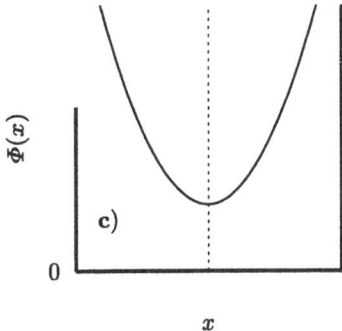

Fig. 4.2a-c. The three qualitative different shapes for the SUSY potential. In case (a) SUSY will be good on the classical as well as on the quantum level. Case (b) displays a situation where SUSY is good on the classical level but will be broken due to quantum fluctuations. The last case (c) shows a SUSY potential where SUSY is already broken explicitly on the classical level.

Thus for a quasi-classical path belonging to class (i), $i = 1, 2, 3, 4$, and containing $k \in \mathbb{N}_0$ complete cycles we obtain [InJu93a, InJu94]

$$\varphi[x_{qc}] \equiv \varphi_k^{(i)} = \varphi_0^{(i)} + k\,[a(x_R) - a(x_L)],\tag{4.39}$$

where

$$\begin{aligned}
\varphi_0^{(1)} &:= \tfrac{1}{2}\,[a(x'') - a(x')],\\
\varphi_0^{(2)} &:= \tfrac{1}{2}\,[a(x'') + a(x')] - a(x_L),\\
\varphi_0^{(3)} &:= a(x_R) - \tfrac{1}{2}\,[a(x'') + a(x')],\\
\varphi_0^{(4)} &:= a(x_R) - a(x_L) - \tfrac{1}{2}\,[a(x'') - a(x')].
\end{aligned}\tag{4.40}$$

The explicit values for $a(x_R)$ and $a(x_L)$ depend on the shape of the SUSY potential Φ. The turning-point condition (4.38) has the two possible solutions

a) $\Phi(x_R) = -\Phi(x_L) = \pm\sqrt{2E}$:
This situation only occurs when SUSY is good on the classical as well as on the quantum level (cf. Fig. 4.2a). We have $a(x_R) = -a(x_L) = \pm\frac{\pi}{2}$ and thus

$$\varphi_k^{(i)} = \varphi_0^{(i)} \pm k\pi.\tag{4.41}$$

b) $\Phi(x_R) = \Phi(x_L) = \pm\sqrt{2E}$:

This is the situation where SUSY is broken on the quantum level. Note that classically SUSY may still be good (cf. Fig. 4.2b and c). Here we find $a(x_R) = a(x_L) = \pm\frac{\pi}{2}$ and consequently

$$\varphi_k^{(i)} = \varphi_0^{(i)}. \tag{4.42}$$

Note that because of the factor two in the exponent of (4.18) the classical fermionic solutions do not require a separate discussion for these two cases. Even more interestingly, they are identical within each class of paths. Denoting with $\psi^{(i)}(t)$ the solution for the class (i) we have

$$\left.\begin{aligned}\psi^{(1)}(t) &= \psi_0 \exp\{-i(a(x'') - a(x'))\} \\ \psi^{(2)}(t) &= -\psi_0 \exp\{-i(a(x'') + a(x'))\} \\ \psi^{(3)}(t) &= -\psi_0 \exp\{i(a(x'') + a(x'))\} \\ \psi^{(4)}(t) &= \psi_0 \exp\{i(a(x'') - a(x'))\}\end{aligned}\right\} = \frac{\psi_0}{2E}\xi^+(0)\xi^-(t) \tag{4.43}$$

and similar expressions for $\bar{\psi}^{(i)}(t)$. See also eq. (4.37).

Here we emphasize that the fermionic phase, in contrast to the fermionic solutions, distinguishes between the cases of good and broken SUSY on the quantum level. Although being a purely classical quantity, the fermionic phase along the quasi-classical path provides some information about the nature of the SUSY of the corresponding quantum system. To be more precise, because of our assumption that $\frac{1}{2}\Phi^2$ forms a single-well potential (cf. Fig. 4.1) the Witten index may be put into the following form [JuMaIn95]

$$\Delta = \frac{1}{\pi}[a(x_R) - a(x_L)]. \tag{4.44}$$

Hence, the fermionic phase accumulates with each complete cycle of the quasi-classical motion an additional phase $\Delta\pi$:

$$\varphi_k^{(i)} = \varphi_0^{(i)} + k\Delta\pi. \tag{4.45}$$

This obviously coincides in the case of good SUSY ($\Delta = \pm1$) with (4.41) and for broken SUSY ($\Delta = 0$) with (4.42).

4.5 Quantization

So far the discussion of this chapter has been devoted to the classical dynamics of supersymmetric systems. Now we want to demonstrate that the quantized version of the model characterized by the Lagrangian (4.10) is identical to the Witten model. There are at least two different approaches for the quantization of a classical dynamical system. These are the standard canonical quantization procedure and the path-integral approach of Feynman [Fey48]. We will sketch both approaches for illustrative purposes. More detailed discussions can be found in [GiPa77b, CoFr83, Nic91, Mat95].

4.5.1 Canonical Approach

Let us begin with the canonical quantization. For this, one first has to derive from Lagrangian (4.10) the corresponding classical Hamiltonian. For this we need the canonical momenta

$$p := \frac{\partial L}{\partial \dot{x}} = \dot{x}, \qquad \pi := \frac{\partial L}{\partial \dot{\psi}} = -\frac{i}{2}\bar{\psi}, \qquad \bar{\pi} := \frac{\partial L}{\partial \dot{\bar{\psi}}} = -\frac{i}{2}\psi. \qquad (4.46)$$

Here we observe that the fermionic momenta π and $\bar{\pi}$ are proportional to the fermionic degrees of freedom. As a consequence, we have the second-class constraints

$$\chi_1 := \pi + \frac{i}{2}\bar{\psi} = 0, \qquad \chi_2 := \bar{\pi} + \frac{i}{2}\psi = 0. \qquad (4.47)$$

This is a typical feature of all pseudoclassical systems [Cas76c]. Obviously, the phase space is six-dimensional. However, the dynamics takes place in a four-dimensional submanifold defined by (4.47). In order to respect these constraints we have to introduce two odd (Grassmann-valued) Lagrange multipliers λ_1, λ_2 and arrive at the so-called total Hamiltonian [Dir64, HeTe92]

$$H_T := \frac{p^2}{2} + \frac{\Phi^2(x)}{2} + \Phi'(x)\bar{\psi}\psi + \left(\pi + \frac{i}{2}\bar{\psi}\right)\lambda_1 + \left(\bar{\pi} + \frac{i}{2}\psi\right)\lambda_2. \qquad (4.48)$$

The corresponding equations of motion read[3]

$$\dot{x} = \frac{\partial H_T}{\partial p}, \qquad \dot{\psi} = -\frac{\partial H_T}{\partial \pi}, \qquad \dot{\bar{\psi}} = -\frac{\partial H_T}{\partial \bar{\pi}},$$
$$\dot{p} = -\frac{\partial H_T}{\partial x}, \qquad \dot{\pi} = -\frac{\partial H_T}{\partial \psi}, \qquad \dot{\bar{\pi}} = -\frac{\partial H_T}{\partial \bar{\psi}}, \qquad (4.49)$$

which are equivalent to the Euler-Lagrange equations derived from L. In fact, using the consistency conditions $\dot{\chi}_1 = 0 = \dot{\chi}_2$ [Dir64, HeTe92], which guarantee that the constraints (4.47) are respected during the time evolution, one finds for the Lagrange multipliers $\lambda_1 = i\Phi'\psi$, $\lambda_2 = \bar{\lambda}_1 = -i\Phi'\bar{\psi}$. Inserting this result in H_T we obtain

$$H_T = \tfrac{1}{2}p^2 + \tfrac{1}{2}\Phi^2(x) + i\Phi'(x)(\bar{\psi}\bar{\pi} - \psi\pi). \qquad (4.50)$$

In this representation the total Hamiltonian obviously gives rise to the correct equations of motion (4.13-4.14). Here let us note that often the classical Hamiltonian

---

[3] Note the extra minus sign for the odd Grassmann variables.

$$H := \tfrac{1}{2} p^2 + \tfrac{1}{2} \Phi^2(x) + \Phi'(x)\bar{\psi}\psi \tag{4.51}$$

is considered, which however, does not lead to the correct equations of motion. Actually, in Dirac's notion H_T and H are *weakly equal*.[4] Such weak equalities are denoted by the sign $\approx$, that is,

$$H \approx H_T. \tag{4.52}$$

However, H and H_T lead to different flows in phase space. It is H_T which generates the correct phase-space flow staying on the submanifold of the phase space defined by the constraints (4.47). Because of these constraints the time evolution of an arbitrary phase-space function is determined by the Dirac bracket instead of the usual Poisson bracket:

$$\frac{\mathrm{d}F}{\mathrm{d}t} \approx \{F, H_T\}_{\mathrm{DB}}, \tag{4.53}$$

where in the present case the Dirac bracket explicitly reads [Mat95]

$$
\begin{aligned}
\{F, G\}_{\mathrm{DB}} = &\frac{\partial F}{\partial x}\frac{\partial G}{\partial p} - \frac{\partial F}{\partial p}\frac{\partial G}{\partial x} \\
&+ (-1)^{\deg F} \frac{1}{2}\left[\frac{\partial F}{\partial \psi}\frac{\partial G}{\partial \pi} + \frac{\partial F}{\partial \pi}\frac{\partial G}{\partial \psi} + \frac{\partial F}{\partial \bar{\psi}}\frac{\partial G}{\partial \bar{\pi}} + \frac{\partial F}{\partial \bar{\pi}}\frac{\partial G}{\partial \bar{\psi}}\right] \\
&+ (-1)^{\deg F} \mathrm{i}\left[\frac{\partial F}{\partial \psi}\frac{\partial G}{\partial \bar{\psi}} + \frac{\partial F}{\partial \bar{\psi}}\frac{\partial G}{\partial \psi} - \frac{1}{4}\frac{\partial F}{\partial \pi}\frac{\partial G}{\partial \bar{\pi}} - \frac{1}{4}\frac{\partial F}{\partial \bar{\pi}}\frac{\partial G}{\partial \pi}\right]
\end{aligned} \tag{4.54}
$$

with $\deg F = 0\ (1)$ if F is an even (odd) Grassmann-valued phase-space function. Note that the Dirac bracket has the following symmetry property:

$$\{F, G\}_{\mathrm{DB}} = \begin{cases} \{G, F\}_{\mathrm{DB}} & \text{for } \deg G = 1 = \deg F \\ -\{G, F\}_{\mathrm{DB}} & \text{else} \end{cases} \tag{4.55}$$

Explicitly we have the following relations

$$
\begin{aligned}
&\{x, p\}_{\mathrm{DB}} = 1, \quad \{\psi, \pi\}_{\mathrm{DB}} = -\tfrac{1}{2} = \{\bar{\psi}, \bar{\pi}\}_{\mathrm{DB}}, \\
&\{\psi, \bar{\psi}\}_{\mathrm{DB}} = -\mathrm{i}, \quad \{\pi, \bar{\pi}\}_{\mathrm{DB}} = \tfrac{\mathrm{i}}{4},
\end{aligned} \tag{4.56}
$$

and all other Dirac brackets for the bosonic and fermionic variables vanish.

Quantization of the system characterized by (4.48) can be achieved[5] by replacing the Grassmann variables by the corresponding operators (for which we will use the same symbol). Simultaneously, the symmetric Dirac brackets are replaced by the (symmetric) anticommutator divided by $(\mathrm{i}\hbar)$. Whereas,

---

[4] Two phase-space functions F and G are called weakly equal, $F \approx G$, if they are identical on the submanifold of the phase space which is defined by the constraints. See ref. [Dir64] p. 12.

[5] See, for example, ref. [HeTe92].

the antisymmetric Dirac brackets are replaced by the (antisymmetric) commutator also divided by $(i\hbar)$. Hence, the operators obey the algebra

$$[x, p] = i\hbar, \quad \{\psi, \pi\} = -\tfrac{i\hbar}{2} = \{\bar\psi, \bar\pi\},$$

$$\{\psi, \bar\psi\} = \hbar, \quad \{\pi, \bar\pi\} = -\tfrac{\hbar}{4}. \qquad (4.57)$$

Note that $\pi^2 = 0 = \bar\pi^2$ and $\psi^2 = 0 = \bar\psi^2$. The algebra satisfied by the fermionic operators is isomorphic to that obeyed by Pauli matrices. Therefore, we may choose the representation[6]

$$\psi = \sqrt{\hbar}\,\sigma_-, \quad \bar\psi = \sqrt{\hbar}\,\sigma_+, \quad \pi = -\frac{i}{2}\sqrt{\hbar}\,\sigma_+, \quad \bar\pi = -\frac{i}{2}\sqrt{\hbar}\,\sigma_-. \qquad (4.58)$$

Using this representation in (4.50) we finally arrive at Witten's quantum mechanical Hamilton operator acting on $L^2(\mathbb{R}) \otimes \mathbb{C}^2$,

$$H_{\mathrm{T}} = \tfrac{1}{2}p^2 + \tfrac{1}{2}\Phi^2(x) + \tfrac{\hbar}{2}\Phi'(x)\sigma_3 \qquad (4.59)$$

with unit mass, $m = 1$, and rescaled SUSY potential $\Phi/\sqrt{2}$. It is also worth mentioning that the ladder operators (3.3) are the quantized versions of the momenta $\xi^\pm$, that is,

$$A = \frac{i}{\sqrt{2}}\,\xi^-, \quad A^\dagger = -\frac{i}{\sqrt{2}}\,\xi^+ \qquad (4.60)$$

and consequently the classical supercharges (4.28) become, after quantization, the supercharges (3.2). Similarly, the relation $E = \tfrac{1}{2}\xi^+\xi^-$ is the classical analogue of the tree Hamiltonian $H_{\mathrm{tree}} = \tfrac{1}{2}\{A^\dagger, A\}$. Let us also mention, that the classical supercharges (4.28) and the classical Hamiltonian (4.48) obey the classical superalgebra

$$\{Q, \bar Q\}_{\mathrm{DB}} \approx -iH_{\mathrm{T}}, \quad \{Q, H_{\mathrm{T}}\}_{\mathrm{DB}} \approx 0, \quad \{\bar Q, H_{\mathrm{T}}\}_{\mathrm{DB}} \approx 0. \qquad (4.61)$$

4.5.2 Path-Integral Approach

The second approach of quantizing the pseudoclassical system (4.10) is based on Feynman's path integral [Fey48, FeHi65]. The basic proposition of Feynman is that the integral kernel of the quantum-mechanical time-evolution operator can be expressed as an integral over a certain set of continuous paths in (pseudo-) classical configuration space, each of which is weighted with a phase given by the action calculated along these paths. For a later convenience, we will only consider the trace of the time evolution operator $\exp\{-itH/\hbar\}$. Clearly, this requires a suitable regularization, for example, by letting the time t have a negative imaginary part, if the Hamiltonian H is bounded from below. In case of a purely classical system described by a standard Lagrangian $L(x, \dot x)$ Feynman's approach to construct H from L can then be written as

---

[6] Note that there are no other (inequivalent) irreducible representations [LaMi89].

$$\mathrm{Tr}\exp\left\{-\frac{\mathrm{i}}{\hbar}tH\right\} = \int\limits_{x(t)=x(0)} \mathcal{D}[x]\exp\left\{\frac{\mathrm{i}}{\hbar}\int_0^t \mathrm{d}\tau\, L(x(\tau),\dot{x}(\tau))\right\}. \qquad (4.62)$$

Here $\int_{x(t)=x(0)} \mathcal{D}[x](\cdot)$ symbolizes integration over all continuous paths x : $[0,t] \to \mathbb{R}$, which are periodic in the sense that $x(t) = x(0)$. More precise definitions of the Feynman path integration can be based, for example, on a theory of so-called pro-measures [AlHo76, DeWiMaNe79, AlBr95, CaDeWi95] or on a discrete time-lattice approach. The latter approach is equivalent to the Lie-Trotter formula [Fey48, Nel64, FeHi65, LeSchm77, Schu81, LaRoTi82, Exn85, InKuGe92, Kle95]. In the following we will perform only formal calculations for simplicity.

The path-integral approach is not only useful for quantizing bosonic degrees of freedom but is also applied to field theories with fermionic degrees of freedom [Ber66]. The formulation of fermionic path integrals is based on the integration rules for Grassmann variables (see, for example, [ZiJu93, Roe94]):

$$\int \mathrm{d}\theta\, 1 = 0, \qquad \int \mathrm{d}\theta\, \theta = 1. \qquad (4.63)$$

From these follows the "Gauss formula"

$$\int \mathrm{d}\theta_1 \int \mathrm{d}\bar{\theta}_1 \cdots \int \mathrm{d}\theta_n \int \mathrm{d}\bar{\theta}_n \exp\left\{\sum_{i,j=1}^n \bar{\theta}_i A_{ij}\theta_j\right\} = \det(A_{ij}) \qquad (4.64)$$

and its infinite-dimensional generalization, the "fermionic Gaussian-path-integral formula"

$$\int\limits_{\psi(t)=\pm\psi(0)} \mathcal{D}[\psi] \int\limits_{\bar{\psi}(t)=\pm\bar{\psi}(0)} \mathcal{D}[\bar{\psi}]\exp\left\{\int_0^t \mathrm{d}\tau \int_0^t \mathrm{d}\tau'\, \bar{\psi}(\tau)D(\tau,\tau')\psi(\tau')\right\}$$
$$= N_\pm \det\left(D(\tau,\tau')\right)_\pm. \qquad (4.65)$$

Again, this path integral is formal in the sense that the determinant on the right-hand side does in general not exist. For this, we have included a "normalization" constant $N_\pm$ to be chosen such that the right-hand side becomes meaningful. We also mention that in the above we consider periodic (+) as well as antiperiodic (−) boundary conditions for the fermionic fields. Clearly, the determinant on the right-hand side does also depend on the choice of the boundary conditions. This dependence is indicated by the subscript. In fact, the natural choice for fermionic degrees of freedom is the antiperiodic one [ZiJu93]. However, we will consider both, periodic and antiperiodic conditions in the following path-integral treatment [GiPa77b, CoFr83]:

$$Z_{\pm}(t) := \int\limits_{x(t)=x(0)} \mathcal{D}[x] \int\limits_{\psi(t)=\pm\psi(0)} \mathcal{D}[\psi] \int\limits_{\bar{\psi}(t)=\pm\bar{\psi}(0)} \mathcal{D}[\bar{\psi}] \exp\left\{\frac{i}{\hbar} \int\limits_0^t d\tau\, L\right\}, \qquad (4.66)$$

where

$$L := \tfrac{1}{2}\dot{x}^2 - \tfrac{1}{2}\Phi^2(x) + i\bar{\psi}\dot{\psi} - \Phi'(x)\bar{\psi}\psi \qquad (4.67)$$

is a Lagrangian equivalent to (4.10) because of $\bar{\psi}(t)\psi(t) = \bar{\psi}(0)\psi(0)$. As the above Lagrangian is bilinear in the fermionic fields, the corresponding path integral is reduced to the problem of calculating the determinant of the kernel

$$D(\tau, \tau') := \left(i\frac{\partial}{\partial\tau} - \Phi'(x(\tau))\right)\delta(\tau - \tau') \qquad (4.68)$$

with appropriate boundary conditions. This problem is immediately reduced to the eigenvalue problem

$$\left(i\frac{\partial}{\partial\tau} - \Phi(x(\tau))\right)\psi_\lambda(\tau) = \lambda\psi_\lambda(\tau) \qquad (4.69)$$

whose solution reads

$$\psi_\lambda(t) = \psi_\lambda(0)\exp\left\{-i\lambda t - i\int\limits_0^t d\tau\,\Phi'(x(\tau))\right\}. \qquad (4.70)$$

From the boundary condition $\psi_\lambda(t) = \pm\psi_\lambda(0)$ one obtains the eigenvalues

$$\lambda_n^- := (2n+1)\frac{\pi}{t} - \frac{1}{t}\int\limits_0^t d\tau\,\Phi'(x(\tau)) \qquad \text{for} \qquad \psi_\lambda(t) = -\psi_\lambda(0),$$

$$\lambda_n^+ := 2n\frac{\pi}{t} - \frac{1}{t}\int\limits_0^t d\tau\,\Phi'(x(\tau)) \qquad \text{for} \qquad \psi_\lambda(t) = +\psi_\lambda(0), \qquad (4.71)$$

where $n \in \mathbb{Z}$. Consequently, the determinant is formally given by the infinite product

$$\det\left(D(\tau, \tau')\right)_\pm = \prod_{n \in \mathbb{Z}} \lambda_n^{\pm}. \qquad (4.72)$$

Defining the normalization constants by

$$\frac{1}{N_-} := 2\prod_{n \in \mathbb{Z}}(2n+1)\frac{\pi}{t}, \qquad \frac{1}{N_+} := 2\prod_{n \in \mathbb{Z}} 2n\frac{\pi}{t} \qquad (4.73)$$

and using the relations

$$\cos z = \prod_{k=1}^\infty\left(1 - \frac{4z^2}{(2k+1)^2\pi^2}\right), \qquad \sin z = z\prod_{k=1}^\infty\left(1 - \frac{z^2}{k^2\pi^2}\right) \qquad (4.74)$$

we arrive at the result

$$Z_\pm(t) = z_-(t) \mp z_+(t), \tag{4.75}$$

where

$$
\begin{aligned}
z_\pm(t) &:= \int\limits_{x(t)=x(0)} \mathcal{D}[x] \exp\left\{ \frac{i}{2\hbar} \int\limits_0^t d\tau \left[\dot{x}^2 - \Phi^2(x) \mp \hbar\Phi'(x) \right] \right\} \\
&= \mathrm{Tr}\, \exp\left\{ -\frac{i}{\hbar} t H_\pm \right\}
\end{aligned}
\tag{4.76}
$$

and

$$H_\pm := \tfrac{1}{2}p^2 + \tfrac{1}{2}\Phi^2(x) \pm \tfrac{\hbar}{2}\Phi'(x) \tag{4.77}$$

are indeed the partner Hamiltonians of the Witten model with unit mass and rescaled SUSY potential $\Phi/\sqrt{2}$. As a result we find that antiperiodic boundary conditions for the fermion degrees of freedom lead to

$$Z_-(t) = \mathrm{Tr}\, \exp\left\{ -\frac{i}{\hbar} t H_T \right\} \tag{4.78}$$

with H_T being the Witten Hamiltonian (4.59). In contrast to this, periodic boundary conditions give rise to

$$Z_+(t) = \mathrm{Tr}\left(W \exp\left\{ -\frac{i}{\hbar} t H_T \right\} \right), \tag{4.79}$$

which is related to the heat-kernel regularized Witten index by

$$Z_+(t) = -\bar{\Delta}(it/\hbar). \tag{4.80}$$

Although this path-integral approach has been formal, it clearly demonstrates that the term proportional to $\hbar$ in the Hamiltonians (4.77) stems from taking all fermion loops into account. This has been the reason for introducing the notion tree Hamiltonian and (fermion-) loop correction in Sect. 3.1.

5. Exact Solution of Eigenvalue Problems

In this chapter we will present some exact results which are obtained via the SUSY formalism. First, we will show that for any standard one-dimensional Schrödinger Hamiltonian one can find a family of associated SUSY potentials [AnBoIo84, Suk85a]. In a second step we will demonstrate that a particular invariance property of such a SUSY potential (the so-called shape invariance [Gen83]) will lead to the exact discrete eigenvalues and their corresponding eigenfunctions of the SUSY Hamiltonian.

5.1 Supersymmetrization of One-Dimensional Systems

In our above discussion on the Witten model we have already seen that several (cf. Sect. 3.5) well-known quantum Hamiltonians possess a SUSY structure according to Definition 2.1.1. Here naturally the question arises: Which quantum systems exhibit a SUSY structure? This question has been considered by de Crombrugghe and Rittenberg [CrRi83] with the general result that SUSY, in particular the extended one with $N > 2$, imposes strong conditions on the Hamiltonian. For one-dimensional one-particle systems, however, it is always possible to bring them into the form of the Witten model [AnBoIo84, Suk85a]. Here we will only consider those systems.

Suppose we are interested in the supersymmetric form of a given quantum problem characterized by the following standard Hamiltonian acting on $L^2(\mathbb{R})$[1]

$$H_V := \frac{p^2}{2m} + V(x), \tag{5.1}$$

where V is some continuous real-valued function such that H_V has a non-empty discrete spectrum which is below a possible continuous spectrum. Suppose further that we want the Hamiltonian H_V to be identified (up to a constant) with the Hamiltonian H_- of a SUSY quantum system of the

---

[1] Without loss of generality we choose as configuration space the Euclidean line $\mathcal{M} = \mathbb{R}$.

Witten type. Note that the other choice H_+ can be treated similar to the procedure described below by replacing the SUSY potential Φ by $-\Phi$. In general the ground-state energy of H_V cannot be expected to vanish or being strictly positive. Hence, in order to be able to accommodate a SUSY structure we have to put

$$H_- := H_V - \varepsilon, \tag{5.2}$$

where ε is an arbitrary constant energy shift sometimes called *factorization energy* [Suk85b]. It is obvious that, if ε equals the ground-state energy of H_V, we will arrive at a good SUSY, whereas for values of ε which are below this ground-state energy we may obtain a broken SUSY. Finally, for ε being above the ground-state energy of H_V the Hamiltonian H_- will have a negative eigenvalue which implies that we cannot get a well-defined SUSY structure.

For the moment let ε be an arbitrary real number. Then the SUSY potential we are looking for is determined by the generalized Riccati equation

$$\Phi^2(x) - \frac{\hbar}{\sqrt{2m}} \Phi'(x) = V(x) - \varepsilon. \tag{5.3}$$

Making the ansatz [Inc56, Suk85b]

$$\begin{aligned}
\Phi_{\varepsilon,\lambda}(x) &= -\frac{\hbar}{\sqrt{2m}} \frac{d}{dx} \ln \left[\rho_\varepsilon(x) \left(1 + \lambda \int_0^x dz\, \rho_\varepsilon^{-2}(z) \right) \right] \\
&= -\frac{\hbar}{\sqrt{2m}} \left(\frac{\rho_\varepsilon'(x)}{\rho_\varepsilon(x)} + \frac{\lambda \rho_\varepsilon^{-2}(x)}{1 + \lambda \int_0^x dz\, \rho_\varepsilon^{-2}(z)} \right)
\end{aligned} \tag{5.4}$$

the non-linear Riccati equation reduces to a linear Schrödinger-like equation:

$$-\frac{\hbar^2}{2m} \rho_\varepsilon''(x) + V(x)\rho_\varepsilon(x) = \varepsilon \rho_\varepsilon(x). \tag{5.5}$$

Note that ρ_ε is not necessarily required to be square integrable. Hence, the parameter ε is indeed arbitrary. Let us remark that with ρ_ε also $\tilde{\rho}_\varepsilon(x) := \rho_\varepsilon(x) \int_0^x dz\, \rho_\varepsilon^{-2}(z)$ is (a linear independent) solution of (5.5) and hence any linear combination $\rho_\varepsilon + \lambda\tilde{\rho}_\varepsilon$, which is used in (5.4), will also yield such a solution.

With $\Phi_{\varepsilon,\lambda}$ as given in (5.4) we have obtained a two-parameter family of SUSY potentials which at least formally brings the original problem (5.1) into a SUSY Hamiltonian H_-. We are also able to find the corresponding partner Hamiltonian

$$H_+ := \frac{p^2}{2m} + \Phi_{\varepsilon,\lambda}^2(x) + \frac{\hbar}{\sqrt{2m}} \Phi_{\varepsilon,\lambda}'(x) = H_V - \varepsilon + \frac{2\hbar}{\sqrt{2m}} \Phi_{\varepsilon,\lambda}'(x), \tag{5.6}$$

which may have energy levels identical to those of H_-. To make it more precise, let us discuss some particular ranges for the value of the parameter ε. For this we will denote the discrete eigenvalues of H_V by ε_n with the ordering $\varepsilon_0 < \varepsilon_1 < \cdots$. We also limit our discussion to the special case $\lambda = 0$ for simplicity.[2] In this case, the zero-energy wave function ϕ_0^- as defined in (3.11) can be identified up to a normalization constant C with ρ_ε:

$$\phi_0^-(x) = C\rho_\varepsilon(x). \tag{5.7}$$

a) $\varepsilon < \varepsilon_0$:

In this case it is well known from Sturmian theory [Inc56] that ρ_ε and therefore ϕ_0^- will have no zeros and will not be normalizable. It is, however, possible that $1/\rho_\varepsilon$ is normalizable. Or in other words, changing the minus sign in front of (5.4) to a plus sign may give rise to a normalizable $\phi_0^+ = C/\rho_\varepsilon$. We are led to case b) discussed below with Φ replaced by $-\Phi$.

In the general case, where neither ρ_ε nor $1/\rho_\varepsilon$ is normalizable, eq. (5.4) gives rise to a well-defined (note that $\lambda = 0$ and ρ_ε has no zeros) SUSY potential with broken SUSY. That is, the partner Hamiltonians H_- and H_+ have identical spectrum. Consequently, the Hamiltonian

$$\tilde{H}_V := H_V + \frac{2\hbar}{\sqrt{2m}}\,\Phi'(x) \tag{5.8}$$

will have the same spectrum as H_V. Thus by supersymmetrization with a factorization energy ε below the ground-state energy of a given Hamiltonian H_V one can find another Hamiltonian $\tilde{H}_V$ with identical spectrum but, of course, different eigenfunctions.

b) $\varepsilon = \varepsilon_0$:

Now $C\rho_\varepsilon = \phi_0^-$ is normalizable and hence, the SUSY potential (5.4) gives rise to a good SUSY. The two partner Hamiltonians H_- and H_+ are essential iso-spectral, that is, the ground-state energy of H_- vanishes and all other eigenvalues of H_- coincide with that of H_+. As a consequence, the Hamiltonian $\tilde{H}_V$, as defined above, has a spectrum which coincides with the set $\{\varepsilon_1, \varepsilon_2, \ldots\}$ of energy eigenvalues of the original Hamiltonian H_V. The SUSY potential is represented by the ground-state wave function [Goz83] of H_-, respectively, H_V:

$$\Phi(x) \equiv \Phi_{\varepsilon_0,0}(x) = -\frac{\hbar}{\sqrt{2m}}\,\frac{(\phi_0^-)'(x)}{\phi_0^-(x)} = -\frac{\hbar}{\sqrt{2m}}\,\frac{\mathrm{d}}{\mathrm{d}x}\ln\phi_0^-(x). \tag{5.9}$$

We will discuss some implications of this relation and the spectral relation between H_V and $\tilde{H}_V$ below.

---

[2] A more general discussion is given by Sukumar [Suk85b].

c) $\varepsilon_{n-1} < \varepsilon \leq \varepsilon_n$, $n = 1, 2, 3, \ldots$:

Here from Sturmian theory we know that ρ_ε has n distinct zeros (of multiplicity one) and hence the SUSY potential (5.4) will have singularities
at those points. Here the superalgebra will only be valid on a formal level
and one has to be careful about a proper definition of the domains for
the operators in question. In particular, the supercharges (3.1) cannot be
self-adjoint because the Hamiltonian $H_- = H_V - \varepsilon$ has n strictly negative
eigenvalues implying complex eigenvalues for the supercharges (3.1). The
formal Hamiltonian H_+ will in general not be essential iso-spectral to H_-
[JeRo84, Fuc86, ShSmVa88].

As an illustrative example let us consider the anharmonic-oscillator potential

$$V(x) := \frac{\hbar^2}{2m} x^n. \tag{5.10}$$

The generalized Riccati equation (5.3) reduces in this case to the usual one if
we set the factorization energy ε to zero, which corresponds to case a) above.
The ansatz (5.4) with $\lambda = 0$ then allows to express ρ_0 in terms of an arbitrary
solution of Bessels differential equation denoted by Z_ν [MaObSo66]:

$$\Phi(x) = -\frac{\hbar}{\sqrt{2m}} \frac{\rho_0'(x)}{\rho_0(x)}, \qquad \rho_0(x) := \sqrt{x}\, Z_{\frac{1}{n+2}}\left(\frac{2i}{n+2}\, x^{(n+2)/2}\right). \tag{5.11}$$

This example shows that a rather simple problem can lead to a complicated
SUSY potential if supersymmetrization according to case a) is chosen. Hence,
the natural choice for the factorization energy is the ground-states energy of
the Hamiltonian (5.1) with the SUSY potential expressed in terms of the
ground-state wave function as in (5.9).

The relation (5.9) displays (for the case of good SUSY) an interesting
relation between the zeros of the SUSY potential and the extrema of the
ground-state wave function. To be explicit, these extrema are precisely given
by the zeros of the SUSY potential. These zeros are in general not identical to
the minima or maxima of the full potential V but coincide with the minima of
Φ^2. This is rather surprising, because one naively expects the maxima (local
minima) of the ground-state wave function to be located at the minima (local
maxima) of the potential V.

Let us consider the following SUSY potential

$$\Phi(x) := \frac{\hbar}{\sqrt{2m}}\, (x - a \tanh x), \qquad a \in \mathbb{R}. \tag{5.12}$$

Obviously, for $a < 1$ this SUSY potential has a single zero at $x_0 = 0$ which is
also the position of the maximum of the ground-state wave function

$$\phi_0^-(x) = C \exp\{-x^2/2\} \cosh^a x. \tag{5.13}$$

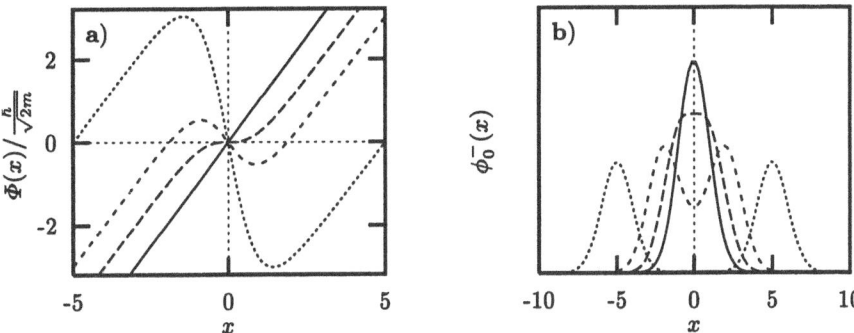

Fig. 5.1. (a) The SUSY potential (5.12) for parameter values $a = 0$ (——), 1 (– – –), 2 (- - - -) and 5 (· · · · ·). **(b)** The ground-state wave function (5.13) for the same parameters as in a).

However, for $a > 1$ the SUSY potential (5.12) has two more zeros, $\Phi(x_{\pm 1}) = 0$, located symmetrically about the first zero $x_0 = 0$ which now becomes the position of a local minimum of ϕ_0^-. The two non-trivial zeros $x_{\pm 1}$ are maxima of the ground-state wave function and are determined by the non-trivial solutions of $x_{\pm 1} = a \tanh(x_{\pm 1})$, $x_{-1} = -x_{+1}$. A graph of the SUSY potential and the ground-state wave function for various values of a are given in Fig. 5.1. The functional dependence of the zeros x_i on the parameter a are shown in Fig. 5.2. It is clearly visible that the single maximum of ϕ_0^- for $a < 1$ bifurcates at $a = 1$ into two maxima at positions $x_{\pm 1}$. In addition the trivial zero $x_0 = 0$ is now the location of the minimum of ϕ_0^-. This bifurcation signals the onset of tunneling in the full potential V_-. The two partner potentials are given by

$$V_\pm(x) = \frac{\hbar^2}{2m} \left[x^2 + a(a \pm 1) \tanh^2 x - 2ax \tanh x \mp (a - 1) \right] \qquad (5.14)$$

and are shown in Fig. 5.3 and 5.4. In particular we note, that $V_-(0) = \frac{\hbar^2}{2m}(a - 1) > 0$ for $a > 1$ but the ground-state energy is identical zero for all values of a due to good SUSY. Hence, $a = 1$ characterizes indeed the onset of tunneling. We also point out that for $a > \frac{1}{2}(3 + \sqrt{5})$ the potential V_- has a local minimum at $x = 0$ where also the ground-state wave function has a local minimum. Therefore, a local minimum of a given potential does not necessary imply a local maximum in the ground-state wave function.

Finally, we note that this example demonstrates that SUSY can also be used for the discussion of tunneling problems. In fact, the tunneling splitting for the lowest energy eigenvalues is given by the lowest eigenvalue of H_+ if SUSY is good. This eigenvalue can, for example, easily be obtained using

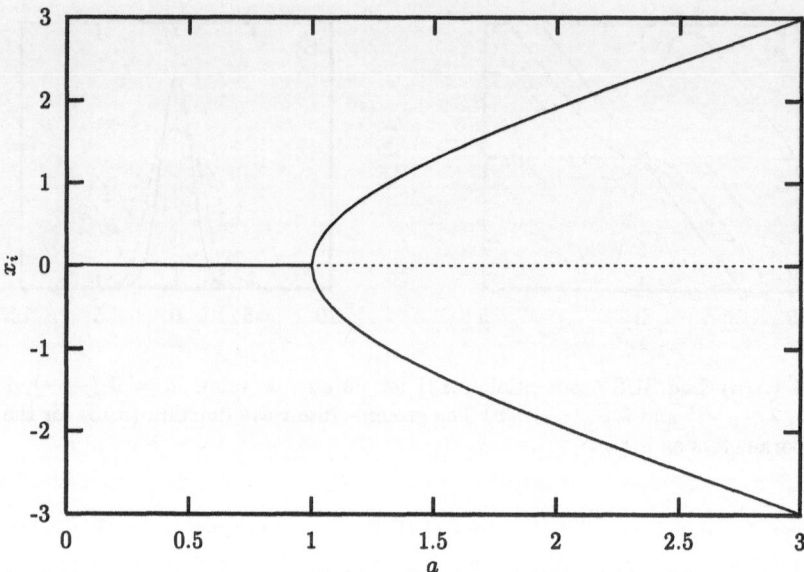

Fig. 5.2. The locations x_i of the maxima (——) and minima ($\cdots\cdots$) of the ground-state wave function (5.13) as a function of the parameter a.

the instanton method [GiPa77a, SaHo82, KaMi89], a perturbative method [SaHo82, KeKoSu88, GaPaSu93] or a WKB-like method [GiKoReMa89].

We have seen in the above discussion that for a factorization energy $\varepsilon = \varepsilon_0$ it is always possible to find for a given Hamiltonian H_V an associated partner Hamiltonian $\widetilde{H}_V$ given in (5.8) which is essential iso-spectral to H_V. It is obvious that by repetition of this supersymmetrization procedure one can construct families of essential iso-spectral Hamiltonians. This method, in fact, has been developed in the last century and is known as Darboux's method [Dar1882].[3] An alternative method, which is based on a theorem of Gel'fand and Levitan [GeLe51, Fad63], has been given by Abraham and Moses [AbMo80]. See also [Suk85a, Suk85b, LuPu86, Pur86]. These methods of constructing a hierarchy of essential iso-spectral Hamiltonians are nowadays successfully applied, for example, in a SUSY variant of the variational [GoReTh93] and inverse scattering method [Suk85c, BaSp94].

5.2 Shape-Invariance and Exact Solutions

In the above discussion we have seen, that, in principle, one can construct a hierarchy of potentials whose corresponding Hamiltonians are pairwise essen-

---

[3] For some historical remarks see [Gro91].

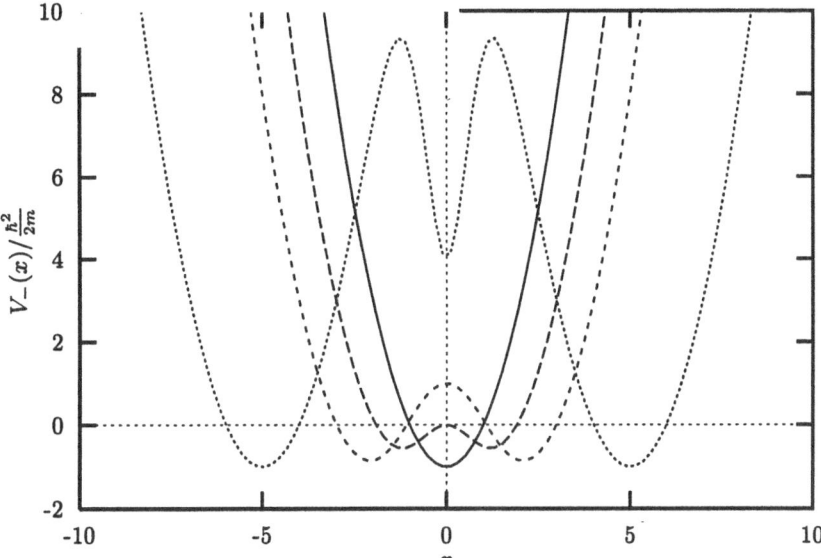

Fig. 5.3. The potential V_- of eq. (5.14) with parameters as in Fig. 5.1.

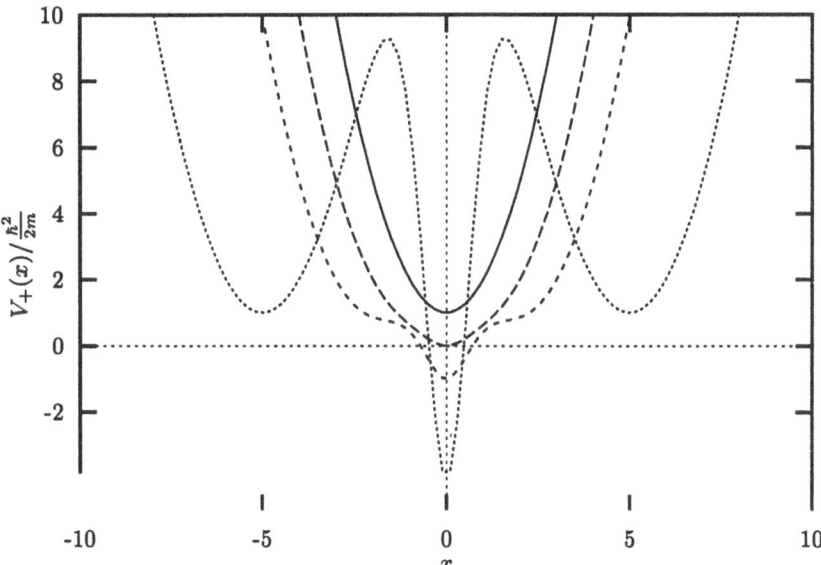

Fig. 5.4. The potential V_+ of eq. (5.14) with parameters as in Fig. 5.1.

tial iso-spectral. In fact, one only needs to know the ground-state energy ε_0 of H_V and the corresponding ground-state wave function ϕ_0^- in order to obtain the partner Hamiltonian $\tilde{H}_V$. Repeating this process for the construction of a hierarchy of pairwise essential iso-spectral Hamiltonians one now needs also the ground-state properties of $\tilde{H}_V$ etc. In other words, in finding a family of such Hamiltonians a complete knowledge of the discrete spectrum of the starting Hamiltonian is required and nothing is gained.

In this section we want to show that under a certain condition called shape-invariance the full information on the discrete spectrum of the starting Hamiltonian is not required and can be obtained in a straightforward way using the SUSY transformation.

Let us assume that we have given a SUSY potential for good SUSY. We further assume that this SUSY potential depends on some set of parameters collectively denoted by a_0. We will explicitly denote the dependence on this parameters in the form $\Phi(a_0, x)$. Then the corresponding partner potentials are given by

$$V_\pm(a_0, x) = \Phi^2(a_0, x) \pm \frac{\hbar}{\sqrt{2m}} \frac{\partial}{\partial x} \Phi(a_0, x). \tag{5.15}$$

Definition 5.2.1 ([Gen83]). *The partner potentials $V_\pm(a_0, x)$ are called shape-invariant if they are related by*

$$\boxed{V_+(a_0, x) = V_-(a_1, x) + R(a_1),} \tag{5.16}$$

where a_1 is a new set of parameters uniquely determined from the old set a_0 via the mapping $F : a_0 \mapsto a_1 = F(a_0)$ and the residual term $R(a_1)$ is independent of the variable x.

In other words, shape-invariance implies that the partner potential $V_+(a_0, x)$ can, after subtracting the constant $R(a_1)$, be interpreted as a new partner potential $V_-(a_1, x)$ associated with a new SUSY potential $\Phi(a_1, x)$.

As an example we mention

$$\Phi(a_0, x) := \frac{\hbar}{\sqrt{2m}} a_0 \tanh x, \qquad a_0 > 0. \tag{5.17}$$

The corresponding partner potentials read

$$V_\pm(a_0, x) = \frac{\hbar^2}{2m} \left[a_0^2 - \frac{a_0(a_0 \mp 1)}{\cosh^2 x} \right] \tag{5.18}$$

and are shape-invariant because of the relation

$$V_+(a_0, x) = V_-(a_0 - 1, x) + \frac{\hbar^2}{2m} \left[a_0^2 - (a_0 - 1)^2 \right]. \tag{5.19}$$

One immediately reads off the function F and R,

$$a_1 = F(a_0) = a_0 - 1, \qquad R(a_1) = \frac{\hbar^2}{2m} \left[a_0^2 - a_1^2 \right], \qquad (5.20)$$

and the ground-state wave function

$$\phi_0^- (a_0, x) = C \cosh^{-a_0} x. \qquad (5.21)$$

Let us now assume, that we have a SUSY potential $\Phi(a_0, x)$ generating shape-invariant partner potentials with a new parameter set a_1 such that $\Phi(a_1, x)$ is also a good SUSY potential. We further assume that the mapping $F : a_{s-1} \mapsto a_s = F(a_{s-1})$ may be iterated n times leading to a family of SUSY potentials $\Phi(a_s, x)$, $s = 0, 1, 2, \ldots, n$, all with good SUSY. Then the discrete energy eigenvalues and corresponding wave functions of the original Hamiltonian $H_0 := \frac{p^2}{2m} + V_-(a_0, x)$ are given by [Gen83, CoGiKh87, RoRoRo91]

$$
\boxed{
\begin{aligned}
E_n &= \sum_{s=1}^{n} R(a_s), \\
\phi_n^- (a_0, x) &= \prod_{s=0}^{n-1} \left(\frac{A^\dagger(a_s)}{[E_n - E_s]^{1/2}} \right) \phi_0^- (a_n, x), \\
\phi_0^- (a_n, x) &= C \exp \left\{ -\frac{\sqrt{2m}}{\hbar} \int\limits_0^x dz\, \Phi(a_n, z) \right\},
\end{aligned}
}
\qquad (5.22)
$$

where the product of the operators

$$A^\dagger(a_s) := -\frac{\hbar}{\sqrt{2m}} \frac{\partial}{\partial x} + \Phi(a_s, x) \qquad (5.23)$$

is ordered such that $A^\dagger(a_s)$ stands to the left of $A^\dagger(a_{s+1})$. The validity of relations (5.22) is an obvious consequence of the SUSY transformation (3.9) which is illustrated in Fig. 5.5. Hence, shape-invariance is a sufficient condition for obtaining complete and exact informations about the spectral properties of the bound-states of the family of Hamiltonians

$$H_s := -\frac{\hbar^2}{2m} \frac{\partial^2}{\partial x^2} + V_-(a_s, x) + E_s, \qquad s = 0, 1, 2, \ldots n. \qquad (5.24)$$

Note that the energy eigenfunctions

$$H_s \phi_{n-s}^- (a_s, x) = E_n \phi_{n-s}^- (a_s, x), \qquad n \geq s, \qquad (5.25)$$

of this family of Hamiltonians are related by

$$\phi_{n-s}^- (a_s, x) = \frac{A^\dagger(a_s)}{\sqrt{E_n - E_s}} \phi_{n-(s+1)}^- (a_{s+1}, x). \qquad (5.26)$$

It should be noted that despite the fact that shape-invariance is a sufficient condition it is not necessary for the solvability of the eigenvalue problem of Schrödinger's equation [CoGiKh87].

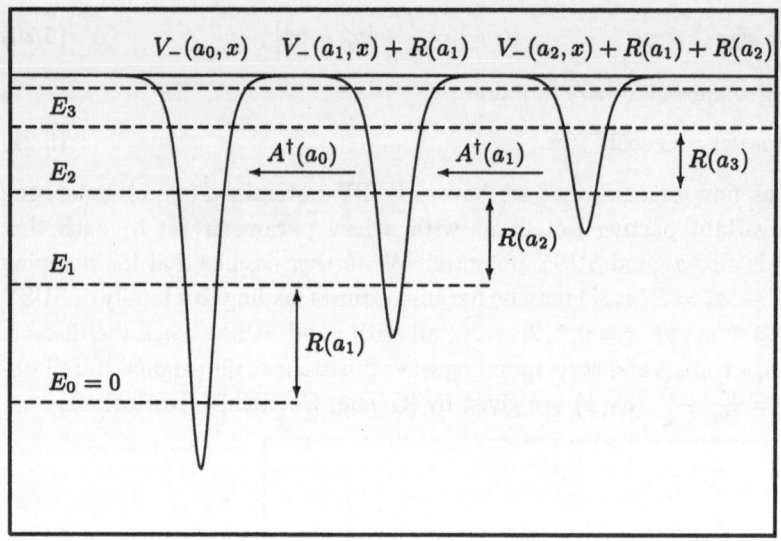

Fig. 5.5. An illustration of a family of shape-invariant potentials. The arrows indicate the relations (5.22) and (5.26).

5.2.1 An Explicit Example

As an example let us briefly discuss the family associated with the SUSY potential (5.17). Note that from (5.20) follows

$$a_s = a_{s-1} - 1 = a_0 - s, \qquad R(a_s) = \frac{\hbar^2}{2m} \left[a_{s-1}^2 - a_s^2 \right], \tag{5.27}$$

and therefore

$$\phi_0^-(a_n, x) = \phi_0^-(a_0 - n, x) = C \cosh^{n-a_0} x \tag{5.28}$$

will be normalizable for all $n \in \mathbb{N}_0$ with $n < a_0$. The eigenvalues are also immediately found:

$$E_n = \frac{\hbar^2}{2m} \sum_{s=1}^{n} (a_{s-1}^2 - a_s^2) = \frac{\hbar^2}{2m} [a_0^2 - (a_0 - n)^2], \tag{5.29}$$

$$n = 0, 1, 2, \ldots < a_0.$$

Finally, the energy eigenfunctions are given by

$$\phi_n^-(a_0, x) = C_n \left(-\frac{\partial}{\partial x} + a_0 \tanh x \right) \times \ldots$$

$$\times \left(-\frac{\partial}{\partial x} + (a_0 - n - 1) \tanh x \right) \cosh^{n-a_0} x, \tag{5.30}$$

where C_n are normalization constants determined by the ground-state normalization C:

$$C_n := C \prod_{s=0}^{n-1} \left[(a_0 - s)^2 - (a_0 - n)^2 \right], \qquad n = 1, 2, 3, \ldots < a_0. \qquad (5.31)$$

5.2.2 Comparison with the Factorization Method

Despite the fact that originally shape-invariance was believed [Gen83] to be a new hidden symmetry which allows for an exact solution of eigenvalue problems, it is equivalent to a similar condition already known from the factorization method [Bau91, Gro91]. The factorization method has been developed by Schrödinger [Schr40, Schr41a, Schr41b] and Infeld and Hull [Inf41, HuIn48, InHu51] as an algebraic method for solving the stationary Schrödinger equation in one dimension [Gre66, deLaRa91]. Its close connection with SUSY quantum mechanics [AnBoIo84, Suk85a] is obvious. In fact, it is amusing to note that already in 1967 Joseph [Jos67], in his search for self-adjoint ladder operators (we would call them now supercharges), found quite naturally that the Hilbert space has to be graded into two subspaces (he called them "a" and "b" subspace). He also noted that the Hamiltonian together with these self-adjoint ladder operators close an algebra containing anticommutators, that is, a superalgebra.

Let us make the connection between factorization method and shape-invariance of SUSY quantum mechanics transparent. Writing the shape-invariance condition in the form

$$\Phi^2(a_{s+1}, x) - \Phi^2(a_s, x) - \frac{\hbar}{\sqrt{2m}} \left[\Phi'(a_{s+1}, x) + \Phi'(a_s, x) \right] = -R(a_{s+1}) \quad (5.32)$$

its equivalence to the condition (3.1.2) of Infeld and Hull [InHu51] becomes obvious if we identify the functions $k(x, s)$ and $L(s)$ in [InHu51] as follows:

$$\Phi(a_s) =: -\frac{\hbar}{\sqrt{2m}} k(x, s), \qquad R(a_{s+1}) =: \frac{\hbar^2}{2m} \left[L(s+1) - L(s) \right]. \qquad (5.33)$$

Note that because of $R(a_{s+1}) > 0$ we arrive at the class I type of factorization of Infeld and Hull. This is due to our convention that the ground state is an eigenstate of H_-. Otherwise, we would arrive at the class II type of factorization. In order to complete our comparison let us note that from (5.33) follows

$$E_n = \sum_{s=1}^{n} R(a_s) = \frac{\hbar^2}{2m} \left[L(n) - L(0) \right], \qquad (5.34)$$

which is identical to Theorem IV of Infeld and Hull ($n = l + 1$), and

$$\phi_0^-(a_n, x) = C \exp \left\{ \int_0^x dz \, k(z, n) \right\},$$ (5.35)

which agrees with eq. (2.7.1) of [InHu51]. Finally, the relation (5.26) coincides up to an overall minus sign with (2.7.2) of [InHu51].

Hence, the shape-invariance condition does not lead to any new insight into the question: Why are some potentials exactly solvable? Shape-invariance is simply a restatement of the factorization condition of Infeld and Hull [InHu51]. In contrast, the solvability of these shape-invariant potentials can be related to an underlying dynamical Lie symmetry [Mil68, OlPe83]. For an embedding of SUSY into those dynamical Lie symmetries see Barut and Roy [BaRo92]. More general dynamical superalgebras are constructed by Baake et al. [BaDeJa91].

The developing interest in SUSY quantum mechanics, however, has revived the important question: Which class of potentials are shape-invariant? This problem was attacked by Infeld and Hull by starting with a typical dependence of the energy eigenvalues on the quantum number n. Recently, Lévai [Lev89] reconsidered this problem in the light of SUSY quantum mechanics. Although his approach was rather different from that of Infeld and Hull, he essentially arrived at the same potentials. See Table 5.1.

However, inspired by this work Wittmer and Weiss [WiWe92] were able to find a shape-invariant generalization of the harmonic-oscillator potential based on the Hongler–Zheng model [HoZh83]. Another new class of shape-invariant potentials has been found by Khare and Sukhatme [KhSu93, BaDuGaKhPaSu93]. Unfortunately, this latter class of potentials cannot be expressed in closed form, that is, in terms of known functions. There exist only series representations which are even not guaranteed to be convergent on the real axis.[4] For a detailed discussion on recent developments in this field we refer to [CoKhSu95].

Finally, for completeness, we also present the classification scheme of shape-invariant potentials as given by Gendenshteı̂n [Gen83]:

---

[4] Actually, these series are believed to have a finite radius of convergence [BaDuGaKhPaSu93]. Hence, the possibility of one or more singularities on the real axis cannot be excluded. In fact, the claimed properties of these potentials (being bounded from below as well as from above because of reflectionlessness and still having an infinite number of bound states) indicates that these potentials do not lead to a Hamiltonian with a well defined domain. In other words, SUSY may only be a formal symmetry algebra. See also our comments on singular SUSY potentials made in Sect. 5.1 case c).

Class 1:

$$\Phi_1(x) := \frac{\hbar}{\sqrt{2m}}\Big(af_1(x) + b\Big) \qquad \text{with}$$

$$f_1'(x) = pf_1^2(x) + qf_1(x) + r, \tag{5.36}$$

$$V_\pm^{(1)}(x) = \frac{\hbar^2}{2m}\left[a(a \pm p)f_1^2(x) + a(2b \pm q)f_1(x) + (b^2 \pm ar)\right].$$

Class 2:

$$\Phi_2(x) := \frac{\hbar}{\sqrt{2m}}\Big(af_2(x) + b/f_2(x)\Big) \qquad \text{with}$$

$$f_2'(x) = pf_2^2(x) + q, \tag{5.37}$$

$$V_\pm^{(2)}(x) = \frac{\hbar^2}{2m}\left[a(a \pm p)f_2^2(x) + \frac{b(b \mp q)}{f_2^2(x)} + 2ab \pm (aq - bp)\right].$$

Class 3:

$$\Phi_3(x) := \frac{\hbar}{\sqrt{2m}}\left(a + b\sqrt{pf_3^2(x) + q}\right)/f_3(x) \qquad \text{with}$$

$$f_3'(x) = \sqrt{pf_3^2(x) + q}, \tag{5.38}$$

$$V_\pm^{(3)}(x) = \frac{\hbar^2}{2m}\left[\frac{a^2 + pq(b \mp 1)}{f_3^2(x)} + \frac{\sqrt{pf_3^2(x) + q}}{f_3^2(x)}a(2b \mp 1) + b^2 p\right].$$

Here $a, b, p, q, r \in \mathbb{R}$ are arbitrary potential parameters. Depending on the values of these parameters SUSY will be good or broken. Let us note that the potentials belonging to class 2 posses the additional reparameterization invariances:

$$\left.\begin{array}{l} a \to -(a \pm p) \\ b \to -(b \mp q) \end{array}\right\} \quad \Rightarrow \quad V_\pm^{(2)}(x) \to V_\pm^{(2)}(x) + \text{const.} \tag{5.39}$$

That is, in both cases the full potentials are only shifted by an additional constant. This reparametrization implies the following changes in the SUSY potential:

$$a \to -(a \pm p) \quad \Rightarrow \quad \Phi_2(x) \to -(a \pm p)f_2(x) + b/f_2(x) \tag{5.40}$$

$$b \to -(b \mp q) \quad \Rightarrow \quad \Phi_2(x) \to af_2(x) - (b \mp q)/f_2(x) \tag{5.41}$$

The particular form of Φ_2 shows that if the parameters originally have been chosen such that SUSY is good, after reparametrization SUSY will be broken [InJu93a, InJu93b, Sup92, InJuSu93]. Examples which belong to this class are the radial harmonic oscillator (Example 1 of Sect. 3.5.2) and the Pöschl–Teller oscillators (Example 3 of Sect. 3.5.2 and Example 2 of Sect. 3.5.3).

Table 5.1. List of known SUSY potentials giving rise to shape-invariant partner potentials

SUSY potential $\Phi(x)/\frac{\hbar}{\sqrt{2m}}$	config. space[a]	parameter range for good SUSY[b]	partner potentials $V_\pm(x)/\frac{\hbar^2}{2m}$
$A\tanh x + B/\cosh x$	$\mathbb{R}$	$A > 0$	$A^2 + \frac{B^2 - A(A\mp1) + B(2A\mp1)\sinh x}{\cosh^2 x}$
$A\coth x - B/\sinh x$	$\mathbb{R}^+$	$B > A > 0$	$A^2 + \frac{B^2 + A(A\mp1) - B(2A\pm1)\cosh x}{\sinh^2 x}$
$-A\cot x + B/\sin x$	$[0, \pi]$	$A > B > 0$	$-A^2 + \frac{B^2 + A(A\pm1) - B(2A\mp1)\cos x}{\sin^2 x}$
$A\tan x - B\cot x$	$[0, \pi/2]$	$A > 0,\ B > 0^c$	$-(A+B)^2 + \frac{A(A\pm1)}{\cos^2 x} + \frac{B(B\pm1)}{\sin^2 x}$
$A\tanh x - B\coth x$	$\mathbb{R}^+$	$A > B > 0^c$	$(A-B)^2 - \frac{A(A\mp1)}{\cosh^2 x} + \frac{B(B\pm1)}{\sinh^2 x}$
$A\tanh x + B/A$	$\mathbb{R}$	$A > B \geq 0$	$A^2 + \frac{B^2}{A^2} - \frac{A(A\mp1)}{\cosh^2 x} + 2B\tanh x$
$-A\coth x + B/A$	$\mathbb{R}^+$	$B > A > 0$	$A^2 + \frac{B^2}{A^2} + \frac{A(A\pm1)}{\sinh^2 x} - 2B\coth x$
$-A\cot x + B/A$	$[0, \pi]$	$A > 0$	$-A^2 + \frac{B^2}{A^2} + \frac{A(A\pm1)}{\sin^2 x} - 2B\cot x$
$Ax - B/x$	$\mathbb{R}^+$	$A > 0,\ B > 0^c$	$-A(2B \mp 1) + A^2 x^2 + \frac{B(B\pm1)}{x^2}$
$-A/x + B/A$	$\mathbb{R}^+$	$A > 0,\ B > 0$	$\frac{B^2}{A^2} - \frac{2B}{x} + \frac{A(A\pm1)}{x^2}$
$-Ae^{-x} + B$	$\mathbb{R}$	$A > 0,\ B > 0$	$B^2 + A^2 e^{-2x} - A(2B \mp 1)e^{-x}$
$Ax + B$	$\mathbb{R}$	$A > 0$	$(Ax + B)^2 \pm A$

[a] For $x \in \mathbb{R}^+$, $x \in [0, \pi/2]$, and $x \in [0, \pi]$ we impose Dirichlet boundary conditions on the wave functions at $x = 0$, $x = 0, \pi/2$, and $x = 0, \pi$, respectively.

[b] With our convention that the ground state is an eigenstate of H_-.

[c] These examples belong to class 2 of Gendensthein and will give rise to a broken SUSY potential if B is replaced by $-B$.

6. Quasi-Classical Path-Integral Approach

In this chapter we will consider a quasi-classical evaluation of the path-integral for the Witten model. In contrast to the usual semi-classical evaluation of the path integral, where one expands the action about the classical paths up to second order, we propose a modified approach by expanding the action about the quasi-classical paths. We arrive in the case of good SUSY at a quantization condition which has previously been suggested by Comtet, Bandrauk and Campbell [CoBaCa85]. For broken SUSY we find a modified form of this quantization condition [InJu93a]. A remarkable property of these two quasi-classical SUSY formulas is that they yield the exact discrete spectrum for all shape-invariant potentials. In combination with the usual WKB formula they are also useful for not shape-invariant (i.e. not exactly soluble) potentials.

6.1 The Path-Integral Formalism

As already mentioned in Sect. 4.5.2, according to Feynman [Fey48, FeHi65] the kernel of the time-evolution operator generated by the standard Schrö-dinger Hamiltonian,

$$H := \frac{p^2}{2m} + V(x), \tag{6.1}$$

is expressible in terms of a path integral

$$\langle x'' | \exp\{-(i/\hbar)tH\} | x' \rangle = \int_{x(0)=x'}^{x(t)=x''} \mathcal{D}[x] \exp\left\{ \frac{i}{\hbar} \int_0^t d\tau\, L\big(x(\tau), \dot{x}(\tau)\big) \right\} \tag{6.2}$$

with the standard Lagrangian

$$L(x, \dot{x}) := \frac{m}{2}\dot{x}^2 - V(x). \tag{6.3}$$

Here, in contrast to the path integral in Sect. 4.5.2, $\int_{x(0)=x'}^{x(t)=x''} \mathcal{D}[x](\cdot)$ symbolizes integration over all continuous paths $x : [0, t] \to \mathbb{R}$ starting in $x' := x(0)$ and ending in $x'' := x(t)$.

From the integral kernel of the time-evolution operator one can obtain the Green function, that is, the integral kernel of the resolvent of H via (complex) Laplace transformation:

$$\langle x''|(E - H)^{-1}|x'\rangle = \frac{1}{i\hbar} \int_0^\infty dt \, \langle x''|e^{-(i/\hbar)tH}|x'\rangle \, e^{(i/\hbar)tE}, \quad \operatorname{Im} E > 0. \quad (6.4)$$

The real poles (in the complex E-plane) of this expression give rise to the discrete spectrum of H. The associated residues provide the corresponding normalized energy eigenfunctions.

6.1.1 The WKB Approximation in the Path Integral

In the semi-classical approximation [Gut67, Schu81, Gut92] one evaluates the path integral (6.2) using the method of stationary phase. That is, one first looks for all classical paths x_{cl} for which the action

$$S[x] := \int_0^t d\tau \left[\frac{m}{2} \dot{x}^2 - V(x) \right] \quad (6.5)$$

is stationary, $\delta S[x_{cl}] = 0$, and which obey $x_{cl}(0) = x'$ and $x_{cl}(t) = x''$. Then one expands the action about these classical paths up to second order in $\eta(\tau) := x(\tau) - x_{cl}(\tau)$,

$$S[x] \simeq S[x_{cl}] + \int_0^t d\tau \left[\frac{m}{2} \dot{\eta}^2 - \frac{1}{2} V''\left(x_{cl}(\tau)\right)\eta^2 \right], \quad (6.6)$$

and thus arrives at

$$\langle x''| \exp\left\{-(i/\hbar)tH\right\} |x'\rangle \simeq \sum_{x_{cl}}^{\text{fixed } t} F_V[x_{cl}] \exp\{(i/\hbar)S[x_{cl}]\} \quad (6.7)$$

with the Fresnel-type path integral

$$F_V[x_{cl}] := \int_{\eta(0)=0}^{\eta(t)=0} \mathcal{D}[\eta] \exp\left\{ \frac{i}{\hbar} \int_0^t d\tau \left[\frac{m}{2} \dot{\eta}^2 - \frac{1}{2} V''\left(x_{cl}(\tau)\right)\eta^2 \right] \right\}. \quad (6.8)$$

In the above the symbol $\sum_{x_{cl}}^{\text{fixed } t}(\cdot)$ stands for the summation over all classical paths starting in x' and arriving after the fixed time t in x''. The remaining path integral (6.8) is easily calculated via the van Vleck-Pauli-Morette formula [Vle28, Pau51, Mor52]

$$
\begin{aligned}
F_V[x_{cl}] &= \sqrt{\frac{i}{2\pi\hbar} \frac{\partial^2 S[x_{cl}]}{\partial x'' \partial x'}} \\
&= \frac{e^{i\pi/4}}{\sqrt{2\pi\hbar}} \left| \frac{\partial^2 S[x_{cl}]}{\partial x'' \partial x'} \right|^{1/2} \exp\{-i\mu[x_{cl}]\pi/2\},
\end{aligned}
\quad (6.9)
$$

where $\mu \in \mathbb{N}_0$ denotes the number of negative eigenvalues of the second variation of the classical action $S[x_{cl}]$. This integer-valued functional μ is usually called Morse index and equals the number of points along x_{cl} which are conjugate to x' [Schu81].

In a second step one performs the Laplace integration (6.4) using, again, the method of stationary phase. A detailed discussion of this calculation can be found in Chap. 18 of Schulman's book [Schu81], which is based on the original work of Gutzwiller [Gut67]. The result reads

$$\langle x''|(E - H)^{-1}|x'\rangle \simeq \frac{1}{i\hbar} \sqrt{D_V(x', x'', E)}$$
$$\times \sum_{x_{cl}}^{\text{fixed } E} \exp\left\{\frac{i}{\hbar} W[x_{cl}] - i\nu[x_{cl}]\frac{\pi}{2}\right\}, \tag{6.10}$$

where

$$D_V(x'', x', E) := \frac{m}{2} |(E - V(x''))(E - V(x'))|^{-1/2} \tag{6.11}$$

and

$$W[x_{cl}] := S[x_{cl}] + Et = \int_{x_{cl}} dx \sqrt{2m(E - V(x))} \tag{6.12}$$

is Hamilton's characteristic functional. Note that $t = t(E)$ is obtained from the relation $\partial S[x_{cl}]/\partial t = -E$ which follows from the stationarity condition in the evaluation of the time integration (6.4). The integral in (6.12) has to be taken along the classical path x_{cl}. In (6.10) the symbol $\sum_{x_{cl}}^{\text{fixed } E}(\cdot)$ denotes the summation over all classical paths from x' to x'' with a given fixed energy E. The integer-valued functional $\nu \in \mathbb{N}_0$ is called Maslov index and equals the number of turning points along the classical path [Gut67, Fel87].

If one finally assumes that the potential V has a single global minimum (cf. Fig. 6.1) the sum over all classical paths in (6.10) can be performed explicitly [Schu81] (see also the next section). We arrive at the well-known quantization condition of Wentzel, Brillouin, and Kramers (WKB formula) [Wen26, Bri26, Kra26, Dun31]:

$$\boxed{\int_{q_L}^{q_R} dx \sqrt{2m(E - V(x))} = \hbar\pi(n + 1/2), \qquad n = 0, 1, 2, \ldots,} \tag{6.13}$$

where q_L and q_R are the classical left and right turning points determined by $V(q_L) = E = V(q_R)$. See also Fig. 6.1. The corresponding approximate energy eigenfunctions read

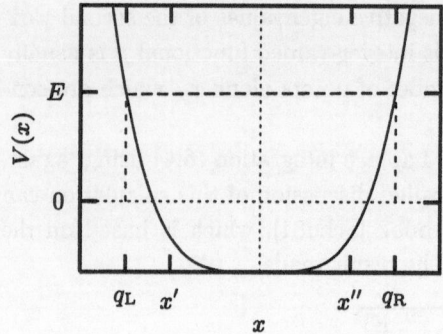

Fig. 6.1. A single-well potential V and the corresponding classical paths for a fixed energy E starting in x' and arriving in x''. The left and right turning points q_L and q_R may be passed several times.

$$\phi_n(x) \simeq \sqrt{\frac{4m}{T_{E_n} p_{E_n}(x)}} \, \sin \left(\frac{1}{\hbar} \int\limits_{q_L}^{x} dz \, p_{E_n}(z) + \frac{\pi}{4} \right), \tag{6.14}$$

where $p_E(x) := \sqrt{2(E - V(x))}$ is the magnitude of the classical momentum for a given energy E, $T_E := 2m \int_{q_L}^{q_R} dx \, [p_E(x)]^{-1}$ is the period of the classical motion and E_n is the solution of the WKB formula (6.13) for a given $n \in \mathbb{N}_0$.

6.1.2 Quasi-Classical Modification for Witten's Model

Let us now consider the case of Witten's model, where the potential appearing in the action (6.5) is given by one of the partner potentials $V_\pm$. That is, the two actions associated with Witten's partner Hamiltonians $H_\pm$ read

$$S^\pm[x] := \int\limits_0^t d\tau \left[\frac{m}{2} \dot{x}^2 - \Phi^2(x) \mp \frac{\hbar}{\sqrt{2m}} \Phi'(x) \right]. \tag{6.15}$$

According to our discussion of Sect. 4.5 we may split this actions into a tree part and a fermion-loop correction,

$$S^\pm[x] = S_{\text{tree}}[x] \mp \hbar\varphi[x], \tag{6.16}$$

where

$$S_{\text{tree}}[x] := \int\limits_0^t d\tau \left[\frac{m}{2} \dot{x}^2 - \Phi^2(x) \right], \qquad \varphi[x] := \frac{1}{\sqrt{2m}} \int\limits_0^t d\tau \, \Phi'(x). \tag{6.17}$$

Note that the above functional φ is identical to the fermionic phase (4.19).[1] The explicit appearance of $\hbar$ in front of the fermionic phase indicates that

---

[1] Note that here we have the original SUSY potential Φ, whereas in (4.19) the rescaled SUSY potential $\Phi/\sqrt{2}$ is used and the mass has been set to unity.

this part of the action stems from quantum corrections of Fermion loops. It is therefore reasonable to assume that the major contributions to the path integral will be supplied by paths which make the tree-action functional stationary. Those paths are in fact the quasi-classical paths introduced in Sect. 4.3. Note, however, that the SUSY potential Φ will in general depend on Planck's constant, too. See, for example (3.12). Actually, the natural units of Φ are $\hbar/\sqrt{2m}$. Hence, a formal power counting as done by Comtet et al. [CoBaCa85] and Eckhardt [Eck86] has no a priori justification. Our suggested modification, which takes the quasi-classical paths and their quadratic fluctuations as dominant contributions to the path integral into account, is not based on such a formal power counting. It rather stems from our discussion of supersymmetric classical dynamics, which has already shown that the solutions of the classical equations of motion are governed by those quasi-classical paths. For this reason the present approach [InJu93a, InJu93b, InJu94] has been called *quasi-classical approximation* in order to distinguish it from the usual semi-classical approximation as discussed in Sect. 6.1.1.

Therefore, let us expand the tree action up to second order in $\eta(\tau) :=$ $x(\tau) - x_{\mathrm{qc}}(\tau)$, that is,

$$S^{\pm}[x] \simeq S_{\mathrm{tree}}[x_{\mathrm{qc}}] \mp \hbar \varphi[x_{\mathrm{qc}}] + \int_0^t \mathrm{d}\tau \left[\frac{m}{2} \dot{\eta}^2 - \frac{1}{2} \left(\Phi^2 \right)'' \left(x_{\mathrm{qc}}(\tau) \right) \eta^2 \right] . (6.18)$$

Then we again arrive at an approximate Fresnel-type path-integral representation for the kernel of the time-evolution operator associated with the SUSY partner Hamiltonians:

$$\langle x'' | e^{-(\mathrm{i}/\hbar)t H_{\pm}} | x' \rangle \simeq \sum_{x_{\mathrm{cl}}}^{\text{fixed } t} F_{\Phi^2}[x_{\mathrm{cl}}] \exp \left\{ \frac{\mathrm{i}}{\hbar} S_{\mathrm{tree}}[x_{\mathrm{qc}}] \mp \mathrm{i}\varphi[x_{\mathrm{qc}}] \right\}, \quad (6.19)$$

where

$$\begin{aligned} F_{\Phi^2}[x_{\mathrm{cl}}] &= \sqrt{\frac{\mathrm{i}}{2\pi\hbar} \frac{\partial^2 S_{\mathrm{tree}}[x_{\mathrm{qc}}]}{\partial x'' \partial x'}} \\ &= \frac{e^{\mathrm{i}\pi/4}}{\sqrt{2\pi\hbar}} \left| \frac{\partial^2 S_{\mathrm{tree}}[x_{\mathrm{qc}}]}{\partial x'' \partial x'} \right|^{1/2} \exp\{-\mathrm{i}\mu[x_{\mathrm{qc}}]\pi/2\}. \end{aligned} \quad (6.20)$$

In essence, the calculation is identical to that of Sect. 6.1.1 with V replaced by Φ^2. The only difference is the additional fermionic phase-functional φ. Hence, we immediately arrive at the approximated Green function

$$\begin{aligned} \langle x'' | (E - H_{\pm})^{-1} | x' \rangle &\simeq \frac{1}{\mathrm{i}\hbar} \sqrt{D_{\Phi^2}(x', x'', E)} \\ &\times \sum_{x_{\mathrm{qc}}}^{\text{fixed } E} \exp \left\{ \frac{\mathrm{i}}{\hbar} W_{\mathrm{tree}}[x_{\mathrm{qc}}] \mp \mathrm{i}\varphi[x_{\mathrm{qc}}] - \mathrm{i}\nu[x_{\mathrm{qc}}]\frac{\pi}{2} \right\} \end{aligned} \quad (6.21)$$

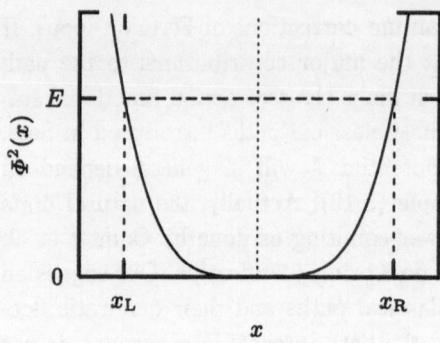

Fig. 6.2. The single-well potential Φ^2 and the left and right turning points x_L and x_R of the quasi-classical motion for a given energy E.

Table 6.1. Phases of the $k = 0$ paths contributing to the resolvent kernel (6.21) within the quasi-classical approximation

Class (i)	$W_0^{(i)}$	$\varphi_0^{(i)}$	$\nu_0^{(i)}$
(1)	$w(x'') - w(x')$	$\frac{1}{2}[a(x'') - a(x')]$	0
(2)	$w(x'') + w(x')$	$\frac{1}{2}[a(x'') + a(x')] - a(x_L)$	1
(3)	$2w(x_R) - w(x'') - w(x')$	$-\frac{1}{2}[a(x'') + a(x')] + a(x_R)$	1
(4)	$2w(x_R) - w(x'') + w(x')$	$-\frac{1}{2}[a(x'') + a(x')] + a(x_R) - a(x_L)$	2

with

$$W_{\text{tree}}[x_{qc}] := \int_{x_{qc}} dx \sqrt{2m(E - \Phi^2(x))}. \tag{6.22}$$

We will now explicitly perform the above path summation. To this end we assume that Φ^2 will have one single global minimum. See Fig. 6.2. Note that this does not necessarily imply that the full potentials $V_\pm$ are of this type, too. In doing this summation we first classify, following Sect. 4.4, the quasi-classical paths into the same four classes (i), $i = 1, 2, 3, 4$. All three phases appearing in (6.21) are functionals of the quasi-classical path. Hence, within each class (i) we can explicitly put them into the form

$$W_{\text{tree}}[x_{cq}] \equiv W_k^{(i)} = W_0^{(i)} + 2kw(x_R),$$

$$\nu[x_{cq}] \equiv \nu_k^{(i)} = \nu_0^{(i)} + 2k, \tag{6.23}$$

$$\varphi[x_{cq}] \equiv \varphi_k^{(i)} = \varphi_0^{(i)} + k[a(x_R) - a(x_L)],$$

where[2]

$$w(x) := \int_{x_L}^{x} dz \sqrt{2m(E - \Phi^2(z))}, \qquad a(x) := \arcsin \frac{\Phi(x)}{\sqrt{E}}. \qquad (6.24)$$

As in Sect. 4.4, the integer $k \in \mathbb{N}_0$ enumerates the number of complete cycles of a path within each class. The quantities for $k = 0$ are given in Table 6.1 with the turning points $x_{L/R}$ to be derived from $\Phi^2(x_L) = E = \Phi^2(x_R)$. The path sum may be rewritten as

$$\sum_{x_{qc}}^{\text{fixed } E} (\cdot) = \sum_{i=1}^{4} \sum_{k=0}^{\infty} (\cdot) \qquad (6.25)$$

and gives rise to a geometric series for the k-part. Performing the latter summation we arrive at the approximate Green function:

$$\langle x'' | (E - H_{\pm})^{-1} | x' \rangle \simeq \frac{1}{i\hbar} \frac{m}{\sqrt{p_E^{qc}(x'') p_E^{qc}(x')}}$$

$$\times \frac{\sum_{i=1}^{4} \exp\left\{ \frac{i}{\hbar} W_0^{(i)} \mp i\varphi_0^{(i)} - i\nu_0^{(i)} \frac{\pi}{2} \right\}}{1 - \exp\left\{ i \left[\frac{2}{\hbar} w(x_R) \mp \left(a(x_R) - a(x_L) \right) - \pi \right] \right\}}, \qquad (6.26)$$

where

$$p_E^{qc}(x) := \sqrt{2m(E - \Phi^2(x))} \qquad (6.27)$$

is the magnitude of the quasi-classical momentum for a given energy E. From this resolvent kernel we may derive approximate energy eigenvalues and eigenfunctions.

6.2 Quasi-Classical Quantization Conditions

From the poles of (6.26) we obtain the following quasi-classical quantization condition

$$w(x_R) \equiv \int_{x_L}^{x_R} dx\, p_E^{qc}(x) = \hbar\pi \left(n + \frac{1}{2} \pm \frac{a(x_R) - a(x_L)}{2\pi} \right), \qquad (6.28)$$

with $n \in \mathbb{N}_0$. This expression can be made more explicit by considering the turning-point condition of the quasi-classical paths, which implies $a(x_{R/L}) = \frac{\pi}{2} \text{sgn} \left[\Phi(x_{R/L}) \right]$:

---

[2] Note that the present definition of $a(x)$ differs from that in (4.35) because of the rescaling of the SUSY potential.

$$\int\limits_{x_L}^{x_R} dx\, p_E^{qc}(x) = \hbar\pi \left(n + \frac{1}{2} \pm \frac{1}{4}\left[\operatorname{sgn}\Phi(x_R) - \operatorname{sgn}\Phi(x_L) \right] \right). \qquad (6.29)$$

This expression has first been obtained by Eckhardt [Eck86] via Maslov theory based, however, on a very strong assumption.[3] The above path-integral approach, which is based on the stationary paths of the tree and not the full action, has first been presented in 1991 [InJu93a]. In the following we will denote the solutions of this quantization condition for a given $n \in \mathbb{N}_0$ by $E_n^{\pm}$ for $H_{\pm}$.

For a discussion of the above quantization condition (6.29) we consider the two possible cases $\Phi(x_L) = -\Phi(x_R)$ and $\Phi(x_L) = \Phi(x_R)$.

Case I. $\Phi(x_L) = -\Phi(x_R) = \pm\sqrt{E}$:
This case corresponds to a good SUSY. Note that we have assumed that Φ^2 is a continuous function with a single global minimum. Hence only the three cases shown in Fig. 4.2 are possible. The present case corresponds to Fig. 4.2a) and implies a good SUSY. We also note that because of our ground-state convention (3.19) only the case $\Phi(x_L) = -\Phi(x_R) = -\sqrt{E}$ may occur. We arrive at the quasi-classical quantization conditions for good SUSY:

$$\int\limits_{x_L}^{x_R} dx\, \sqrt{2m(E - \Phi^2(x))} = \hbar\pi n \qquad \text{for } H_-,$$

$$\int\limits_{x_L}^{x_R} dx\, \sqrt{2m(E - \Phi^2(x))} = \hbar\pi(n+1) \qquad \text{for } H_+. \qquad (6.30)$$

Formula (6.30), which is sometimes referred to as CBC formula, has first been suggested by Comtet, Bandrauk and Campbell [CoBaCa85] based on a formal WKB approach. They have assumed that Φ will be independent of $\hbar$.

Comparing the CBC formula (6.30) with the WKB formula (6.13) we note two differences. First, instead of the full potentials $V_{\pm}$ only Φ^2 appears in the integral on the left-hand side. Secondly, the extra term $\frac{1}{2}$ in the WKB formula, which stems from the Maslov indices, is in the case of H_- precisely canceled by the contributions of the fermionic phase. In the case of H_+, however, these two contributions are equal and add up to unity. The CBC formula (6.30) has some remarkable properties:

---

[3] Eckhardt explicitly assumes that the $\hbar$ dependence of the SUSY potential Φ is such that $\lim_{\hbar \to 0} \Phi^2$ is well-defined. This, for example, is not the case for the radial hydrogen atom (3.47). In fact, for the natural choice of units $\Phi(x) = \frac{\hbar}{\sqrt{2m}} f(x)$, where f is independent of $\hbar$, Eckhardt's assumption covers the free-particle case only.

1) The quantization condition (6.30) provides the exact ground-state energy for H_-. Note that for $n = 0$ one necessarily is led to $E \equiv E_0 = 0$. We have already shown in Sect. 5.1, that the knowledge of the ground-state energy and wave function for a given Hamiltonian is needed in order to find the SUSY potential. Hence, the quasi-classical quantization condition cannot provide an additional information about the ground state. Nevertheless, it is remarkable that the quasi-classical approach can reproduce the exact ground-state energy for good SUSY. This is a rather unexpected result.

2) The presence of the fermionic phase has the consequence that the exact relation $E_{n+1}^- = E_n^+$ is also valid within the quasi-classical approximation. Note that the WKB formula (6.13) does in general not reproduce this relation.

3) An immediate and obvious consequence of properties 1) and 2) is, that the CBC formula will reproduce exact bound-state spectra for all shape-invariant potentials [CoBaCa85, DuKhSu86]. This is in contrast to the WKB formula, which is not able to reproduce the exact spectra for those systems without any ad hoc modifications of the Langer type [RoKr68]. Only in the special case of the harmonic oscillator and the Morse oscillator the WKB formula is able to reproduce the exact bound state spectrum. See also our discussion in Sect. 6.4.1 below. Furthermore, it has been shown that for a wide class of shape-invariant potentials all higher-order corrections (in $\hbar$) to the CBC formula vanish [RaSeVa87, BaMa91].

Case II. $\Phi(x_L) = \Phi(x_R) = \pm\sqrt{E}$:
This is the case where SUSY is broken (at the quantum level). See Figs. 4.2 b) and c). Here the contribution to the fermionic phase from the left turning point cancels that from the right turning point. The quantization condition (6.29) reduces to

$$\int_{x_L}^{x_R} dx \sqrt{2m(E - \Phi^2(x))} = \hbar\pi \left(n + \frac{1}{2}\right) \qquad \text{for } H_\pm. \qquad (6.31)$$

This formula and its relation to broken SUSY have first been discussed in [InJu93a, InJu93b]. As in (6.30), instead of the full potential $V_\pm$ only Φ^2 appears in the integral on the left-hand side. Whereas, on the right-hand side it is identical to the WKB formula. The Maslov indices, are not affected by the fermionic phase. Formula (6.31) has also some remarkable properties:

1) The exact relation $E_n^- = E_n^+$ valid for broken SUSY is respected also by the quasi-classical expression (6.31). Again we note that the WKB formula (6.13) does in general not obey this relation.

2) The surprising result, which has first been noted in [InJu93a] is, that for those shape-invariant potentials for which the parameters can be cho-

sen such that SUSY will be broken (see remark c) in Table 4.1), formula (6.31) does provide the exact bound-state spectrum. See also ref. [InJu93b, InJuSu93, KoSuIn94] and our discussion in Sect. 6.4.1. As for the CBC formula it can be shown that higher-order corrections to (6.31) vanish identical for some shape-invariant potentials [MuGoKh95].

It is interesting to note that (6.30) and (6.31) can be combined with the help of the Witten index (2.43), which in the case of the Witten model is given by $\Delta = \mp 1$ for good SUSY if the ground state belongs to $H_\pm$, and by $\Delta = 0$ for broken SUSY, respectively. See eq. (3.29). Hence, we may write (6.30) and (6.31) in the combined form

$$\boxed{\int\limits_{x_L}^{x_R} dx\, \sqrt{2m(E - \Phi^2(x))} = \hbar\pi\left(n + \frac{1}{2} \pm \frac{\Delta}{2}\right) \qquad \text{for } H_\pm,} \qquad (6.32)$$

which is even independent of any convention about the ground state. The above expression displays an interesting interplay between the Witten and the Maslov index. We will refer to expression (6.32) as the *quasi-classical SUSY (qc-SUSY) approximation* in the following. The appearance of Witten's index in this formula is rather natural. Note that under the assumption we have made, that is, Φ^2 characterizes a single-well potential (cf. Fig. 6.2) we may put expression (3.29) into the form

$$\Delta = \frac{1}{\pi}\Big(a(x_R) - a(x_L)\Big), \qquad x_R > x_L. \qquad (6.33)$$

This identification allows us to interpret the qc-SUSY approximation (6.32) as the pseudoclassical analogue of the Bohr–Sommerfeld quantization condition [JuMaIn95]. In fact, let us consider the following pseudoclassical phase integral along one period of the quasi-classical motion (cf. eqs. (4.15) and (4.46))

$$\oint \left(\pi d\psi + \bar{\pi}d\bar{\psi}\right) = -\left(a(x_R) - a(x_L)\right)[\bar{\psi}_0, \psi_0] = -\pi\Delta[\bar{\psi}_0, \psi_0] \qquad (6.34)$$

and note that $[\bar{\psi}_0, \psi_0]$ is replaced by $\hbar\sigma_3$ upon quantization (cf. eq. (4.58)). Since, σ_3 has eigenvalues ± 1 in the subspace $\mathcal{H}^\pm$ the qc-SUSY formula (6.32) may formally be put into the form

$$\oint \left(p_E^{qc}dx + \pi d\psi + \bar{\pi}d\bar{\psi}\right) = 2\pi\hbar\left(n + \tfrac{1}{2}\right). \qquad (6.35)$$

Let us now comment on the formal approaches of [CoBaCa85, Eck86]. In fact, assuming that Φ is independent of $\hbar$, the above expression (6.32) can be derived from the standard WKB formula (6.13) by simple Taylor expansion in $\hbar$ [InJuSu93]. Replacing in the WKB formula (6.13) the potential V by the explicit form of $V_\pm$ we find:

$$
\int\limits_{q_{\mathrm{L}}}^{q_{\mathrm{R}}} \mathrm{d}x \sqrt{\left(p_E^{\mathrm{qc}}(x)\right)^2 \mp \frac{\hbar}{\sqrt{2m}}\Phi'(x)}
$$

$$
= \int\limits_{x_{\mathrm{L}}}^{x_{\mathrm{R}}} \mathrm{d}x\, p_E^{\mathrm{qc}}(x) \pm \frac{\hbar}{2} \int\limits_{x_{\mathrm{L}}}^{x_{\mathrm{R}}} \mathrm{d}x \frac{\Phi'(x)}{\sqrt{E - \Phi^2(x)}} + O(\hbar^{3/2}),
\tag{6.36}
$$

where we have expanded the square root on the left-hand side in a Taylor series and kept only the first two terms. We have also replaced the classical turning points $V(q_{\mathrm{L/R}}) = E$ by the quasi-classical turning points $\Phi^2(x_{\mathrm{L/R}}) = E$. The second integral on the right-hand side can explicitly be evaluated and we will arrive at the expression (6.28) which is equivalent to (6.32). Hence, for a SUSY potential which is independent of $\hbar$ the quasi-classical SUSY approximation (6.32) is actually equivalent to the WKB formula in first order of $\hbar$. An example for such a case is the harmonic oscillator SUSY potential $\Phi(x) = \sqrt{m/2}\,\omega x$ for which the WKB as well as the CBC formula provide the exact spectrum. This, however, is not a typical but rather an exceptional case. Actually, whenever there is an intrinsic length scale in the underlying classical problem (for example $1/\alpha$ in $\Phi(x) = \hbar a/\sqrt{2m}\,\tanh(\alpha x)$) then the energy has the unit $\hbar^2/2m$ if we measure distances in units of this length scale (i.e. $\alpha = 1$). Setting $\Phi(x) =: \frac{\hbar}{\sqrt{2m}} f(x)$ and $E =: \frac{\hbar^2}{2m}\varepsilon$ the stationary Schrödinger equation reads

$$
\left(-\frac{\partial^2}{\partial x^2} + f^2(x) \pm f'(x)\right)\varphi(x) = \varepsilon\varphi(x).
\tag{6.37}
$$

The WKB and the quasi-classical SUSY formula lead to the following approximate quantization conditions for the dimensionless eigenvalues ε, respectively.

$$
\int\limits_{q_{\mathrm{L}}}^{q_{\mathrm{R}}} \mathrm{d}x \sqrt{\varepsilon - f^2(x) \mp f'(x)} = \pi\left(n + \frac{1}{2}\right),
$$

$$
\int\limits_{x_{\mathrm{L}}}^{x_{\mathrm{R}}} \mathrm{d}x \sqrt{\varepsilon - f^2(x)} = \pi\left(n + \frac{1}{2} \pm \frac{\Delta}{2}\right).
\tag{6.38}
$$

Hence, the WKB and the quasi-classical SUSY approximations are indeed not equivalent.

6.3 Quasi-Classical Eigenfunctions

The approximate result for the resolvent kernel (6.26) also provides the quasi-classical wave functions to be obtained from the residues of its poles [InJu93a]. Again we will distinguish the two cases of good and broken SUSY.

Case I. Good SUSY:

Using the explicit forms of the quantities in Table 5.1 we can perform the remaining sum in (6.26) at the poles $E_n^{\pm}$ determined by the CBC formula (6.30):

$$\left. \mathrm{Res}\langle x''|(E-H_+)^{-1}|x'\rangle \right|_{E=E_n^+}$$

$$\simeq \frac{4m}{T_E^{\mathrm{qc}}} \frac{\sin\left(\frac{w(x')}{\hbar} - \frac{a(x')}{2}\right)}{\sqrt{p_E^{\mathrm{qc}}(x')}} \left. \frac{\sin\left(\frac{w(x'')}{\hbar} - \frac{a(x'')}{2}\right)}{\sqrt{p_E^{\mathrm{qc}}(x'')}} \right|_{E=E_n^+} ,$$

$$\left. \mathrm{Res}\langle x''|(E-H_-)^{-1}|x'\rangle \right|_{E=E_n^-}$$

$$\simeq \frac{4m}{T_E^{\mathrm{qc}}} \frac{\cos\left(\frac{w(x')}{\hbar} + \frac{a(x')}{2}\right)}{\sqrt{p_E^{\mathrm{qc}}(x')}} \left. \frac{\cos\left(\frac{w(x'')}{\hbar} + \frac{a(x'')}{2}\right)}{\sqrt{p_E^{\mathrm{qc}}(x'')}} \right|_{E=E_n^-} ,$$

where $T_E^{\mathrm{qc}} := 2m \int_{x_L}^{x_R} dz \, [p_E^{\mathrm{qc}}(z)]^{-1}$ is the period of the bounded quasi-classical motion for a given energy E. From these residues we read off the quasi-classical wave functions:

$$\boxed{\phi_n^+(x) \simeq \sqrt{\frac{4m}{T_{E_n^+}^{\mathrm{qc}} p_{E_n^+}^{\mathrm{qc}}(x)}} \, \sin\left(\frac{1}{\hbar} \int_{x_L}^{x} dz \, p_{E_n^+}^{\mathrm{qc}}(z) - \frac{1}{2}\arcsin\frac{\Phi(x)}{\sqrt{E_n^+}}\right) ,} \quad (6.40)$$

$$\boxed{\phi_n^-(x) \simeq \sqrt{\frac{4m}{T_{E_n^-}^{\mathrm{qc}} p_{E_n^-}^{\mathrm{qc}}(x)}} \, \cos\left(\frac{1}{\hbar} \int_{x_L}^{x} dz \, p_{E_n^-}^{\mathrm{qc}}(z) + \frac{1}{2}\arcsin\frac{\Phi(x)}{\sqrt{E_n^-}}\right) .} \quad (6.41)$$

Note that $E_0^- = 0$ and, therefore, also $p_{E_0^-} = 0$. Hence, (6.41) is not defined for $n = 0$. We have to exclude the value $n = 0$ in (6.41). Actually, the ground-state wave function is already known exactly

$$\phi_0^-(x) = C \exp\left\{-\frac{\sqrt{2m}}{\hbar} \int_{x_0}^{x} dz \, \Phi(z)\right\} . \quad (6.42)$$

In comparing the expressions (6.40) and (6.41) with the WKB result (6.14) we note that the latter has a constant phase shift $\pi/4$ whereas the SUSY wave functions contain x-dependent phase shifts stemming from the fermionic phase.

Case II. Broken SUSY:

In this case the residues of (6.26) explicitly read

$$\text{Res}\langle x''|(E - H_+)^{-1}|x'\rangle\Big|_{E=E_n^+}$$

$$\simeq \frac{4m}{T_E^{qc}} \frac{\cos\left(\frac{w(x')}{\hbar} - \frac{a(x')}{2}\right)}{\sqrt{p_E^{qc}(x')}} \frac{\cos\left(\frac{w(x'')}{\hbar} - \frac{a(x'')}{2}\right)}{\sqrt{p_E^{qc}(x'')}}\Bigg|_{E=E_n^+} ,$$

$$\text{Res}\langle x''|(E - H_-)^{-1}|x'\rangle\Big|_{E=E_n^-}$$ (6.43)

$$\simeq \frac{4m}{T_E^{qc}} \frac{\sin\left(\frac{w(x')}{\hbar} + \frac{a(x')}{2}\right)}{\sqrt{p_E^{qc}(x')}} \frac{\sin\left(\frac{w(x'')}{\hbar} + \frac{a(x'')}{2}\right)}{\sqrt{p_E^{qc}(x'')}}\Bigg|_{E=E_n^-} ,$$

where now the poles at $E = E_n^+ = E_n^-$ are determined by (6.31). Hence the corresponding quasi-classical wave functions read

$$\boxed{\phi_n^+(x) \simeq \sqrt{\frac{4m}{T_{E_n^+}^{qc} p_{E_n^+}^{qc}(x)}} \cos\left(\frac{1}{\hbar}\int_{x_L}^x dz\, p_{E_n^+}^{qc}(z) - \frac{1}{2}\arcsin\frac{\Phi(x)}{\sqrt{E_n^+}}\right),}$$ (6.44)

$$\boxed{\phi_n^-(x) \simeq \sqrt{\frac{4m}{T_{E_n^-}^{qc} p_{E_n^-}^{qc}(x)}} \sin\left(\frac{1}{\hbar}\int_{x_L}^x dz\, p_{E_n^-}^{qc}(z) + \frac{1}{2}\arcsin\frac{\Phi(x)}{\sqrt{E_n^-}}\right).}$$ (6.45)

As in the good-SUSY case we observe the additional x-dependent phase appearing in (6.44) and (6.45).

The wave functions as they stand (including the WKB wave function) are only valid in the quasi-classical (classical) allowed region $x_L < x < x_R$ ($q_L < x < q_R$). In the forbidden regions the trigonometric functions have to be replaced by appropriate decreasing exponential functions. As expected, these quasi-classical eigenfunctions are singular at the quasi-classical turning points x_L and x_R. For a discussion of regularized wave functions see [FrBaHaUz88, Mur89, PaSu90].

6.4 Discussion of the Results

6.4.1 Exactly Soluble Examples

As we have already mentioned above, the quasi-classical SUSY quantization condition (6.32) provides the exact bound-state eigenvalues for all shape-invariant potentials. This is in contrast to the WKB formula (6.13), which in general requires an ad hoc replacement of the potential parameters in order to yield the exact eigenvalues for those shape-invariant potentials. As an example, let us consider the radial harmonic oscillator

$$V(r) = \frac{m}{2}\omega^2 r^2 + \frac{\hbar^2 l(l+1)}{2mr^2}, \qquad l = 0, 1, 2, \cdots, \qquad r > 0. \tag{6.46}$$

Using the WKB approximation, it is known [RoKr68] that only upon the Langer modification $l(l+1) \to (l+\frac{1}{2})^2$ the WKB formula

$$\int_{q_L}^{q_R} dr \sqrt{2m \left(\tilde{E} - \frac{m}{2}\omega^2 r^2 - \frac{\hbar^2 (l+\frac{1}{2})^2}{2mr^2} \right)} = \hbar\pi(n+1/2) \tag{6.47}$$

will give rise to the exact spectrum

$$\tilde{E}_n = \hbar\omega(2n + l + 3/2). \tag{6.48}$$

Let us now consider the SUSY potential

$$\Phi(r) := \sqrt{\frac{m}{2}}\,\omega r - \frac{\hbar(l+1)}{\sqrt{2m}\,r}, \tag{6.49}$$

which leads to a good SUSY. The associated partner potentials are

$$\begin{aligned} V_+(r) &= \frac{m}{2}\omega^2 r^2 + \frac{\hbar^2 (l+1)(l+2)}{2mr^2} - \hbar\omega(l+1/2), \\ V_-(r) &= \frac{m}{2}\omega^2 r^2 + \frac{\hbar^2 l(l+1)}{2mr^2} - \hbar\omega(l+3/2), \end{aligned} \tag{6.50}$$

where the latter is identical to the original potential V up to a constant negative shift by the ground-state energy $\tilde{E}_0 = \hbar\omega(l+3/2)$. V_+ corresponds to the same potential but with l replaced by $l+1$ and a different energy shift. Using the qc-SUSY formula (6.32)

$$\int_{x_L}^{x_R} dr \sqrt{2m \left(E - \frac{m}{2}\omega^2 r^2 - \frac{\hbar^2 (l+1)^2}{2mr^2} + \hbar\omega(l+1) \right)}$$
$$= \hbar\pi \left(n + \frac{1}{2} \pm \frac{1}{2} \right) \tag{6.51}$$

we immediately arrive at

$$E_n^\pm = 2\hbar\omega \left(n + \frac{1}{2} \pm \frac{1}{2} \right), \tag{6.52}$$

which is the exact result as $\tilde{E}_n = E_n^- + \hbar\omega(l+3/2)$.

On the contrary, choosing the SUSY potential to

$$\Phi(r) := \sqrt{\frac{m}{2}}\,\omega r + \frac{\hbar l}{\sqrt{2m}\,r} \tag{6.53}$$

we realize a broken SUSY with partner potentials

$$V_+(r) = \frac{m}{2}\,\omega^2 r^2 + \frac{\hbar^2(l-1)l}{2mr^2} + \hbar\omega(l+1/2),$$

$$V_-(r) = \frac{m}{2}\,\omega^2 r^2 + \frac{\hbar^2 l(l+1)}{2mr^2} + \hbar\omega(l-1/2). \tag{6.54}$$

Here the latter is also identical to V with a positive shift by $\hbar\omega(l-1/2)$ and V_+ is similar to V_- with l replaced by $l-1$. In this case the qc-SUSY quantization condition reads

$$\int_{x_L}^{x_R} dr \sqrt{2m\left(E - \frac{m}{2}\omega^2 r^2 - \frac{\hbar^2 l^2}{2mr^2} - \hbar\omega l\right)} = \hbar\pi\left(n + \frac{1}{2}\right) \tag{6.55}$$

and yields the energy eigenvalues

$$E_n^\pm = \hbar\omega\left(2n + 2l + 1\right), \tag{6.56}$$

which again are exact because of $\widetilde{E}_n = E_n^- - \hbar\omega(l-1/2)$.

Similar to this example it is straightforward but tedious [Sup92, KoSuIn94] to show that all of the shape-invariant potentials listed in Table 5.1 give rise to the exact energy eigenvalues if the qc-SUSY formula (6.32) is used. For those systems where SUSY is good this phenomenon can be explained according to our remark 3) on page 75.[4] However, for those cases (marked with c) in Table 5.1) which also allow for a broken SUSY potential, there is no explanation available for the exactness of formula (6.31). In fact, the question – why is the qc-SUSY approximation exact for shape-invariant potentials? – is still open. A possible explanation might be based on the discussion of the Nicolai map by Ezawa and Klauder [EzKl85], see also Sect. 7.3 below. Ezawa and Klauder have shown that the (Euclidean-time) path integral for SUSY quantum mechanics can be transformed into a Gaussian path integral via this Nicolai map. Gaussian path integrals, in turn, are exactly evaluated by the method of stationary phase.

6.4.2 Numerical Investigations

We have seen that the qc-SUSY approximation (6.32) does lead to exact bound-state spectra for (by other means) exactly solvable problems. Here immediately comes the question to mind: Does this approximation yield also better eigenvalues (than the WKB approximation (6.13) does) for those problems which are not exactly solvable? To answer this questions we have investigated a class of power SUSY potentials of the form [InJu94]

---

[4] Note that recently it has been argued [BaKhSu93] that for the new shape-invariant potentials found by [KhSu93] the qc-SUSY does not yield the exact eigenvalues. In contrast to the arguments given in [BaKhSu93] this might happen because of the singular nature of these potentials. See the footnote on p. 64.

$$\Phi(x) := \frac{\hbar a}{\sqrt{2m}} x^d \qquad \text{for} \quad x \geq 0. \tag{6.57}$$

Here $a > 0$ and $d \geq 1$ are free parameters. Note that the above definition is only valid for the positive Euclidean half-line. For $x < 0$ we may define the SUSY potential either through an antisymmetric or symmetric continuation, which leads to a good and broken SUSY, respectively:

$$\Phi(x) = \frac{\hbar a}{\sqrt{2m}} |x|^d \operatorname{sgn}(x) \qquad \text{for good SUSY}, \tag{6.58}$$

$$\Phi(x) = \frac{\hbar a}{\sqrt{2m}} |x|^d \qquad \text{for broken SUSY}. \tag{6.59}$$

The associated partner potentials read

$$V_\pm(x) = \frac{\hbar^2}{2m} \left(a^2 x^{2d} \pm a d |x|^{d-1} \right) = V_\pm(-x) \qquad \text{for good SUSY}, \tag{6.60}$$

$$V_\pm(x) = \frac{\hbar^2}{2m} \left(a^2 x^{2d} \pm a d |x|^{d-1} \operatorname{sgn} x \right) = V_\mp(-x) \tag{6.61}$$
$$\text{for broken SUSY}$$

and are shown in Fig. 6.3 and Fig. 6.4, respectively.

For the SUSY potentials (6.58–6.59) the qc-SUSY quantization condition (6.32) can be evaluated analytically with the result (Γ denotes Euler's gamma function)

$$E_n^\pm = \frac{\hbar^2 a^2}{2m} \left[\frac{\Gamma\left(\frac{3d+1}{2d}\right)}{\Gamma\left(\frac{2d+1}{2d}\right)} \frac{\sqrt{\pi}}{a} \left(n + \frac{1}{2} \pm \frac{\Delta}{2} \right) \right]^{2d/(d+1)}, \tag{6.62}$$

where $\Delta = 1$ for the case of good SUSY and $\Delta = 0$ for broken SUSY,

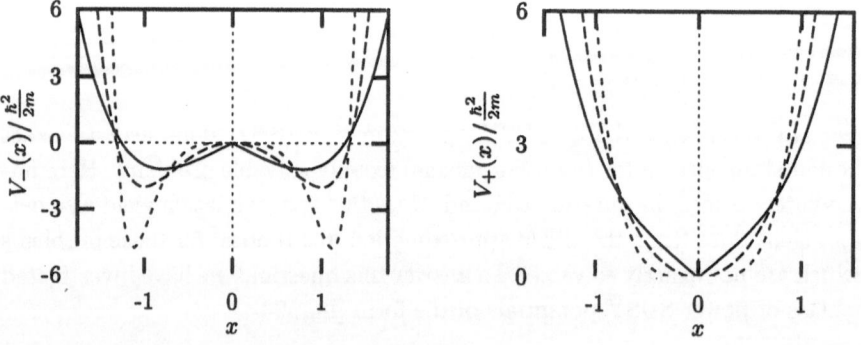

Fig. 6.3. The partner potentials (6.60) for values $a = 1$ and $d = 2$ (*solid line*), 3 (*long dashes*), and 5 (*short dashes*).

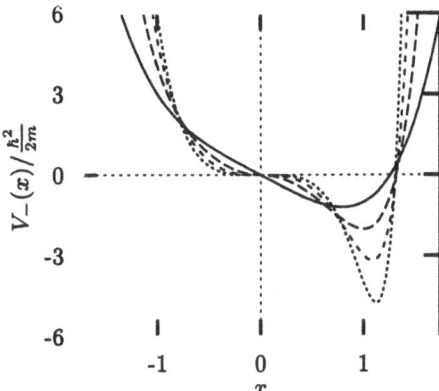

Fig. 6.4. The potential V_- for broken SUSY (6.61) and $a = 1$, $d = 2$ (*solid line*), 3 (*long dashes*), 4 (*short dashes*), and 6 (*dotted line*).

respectively. These energy eigenvalues are in general not exact. For arbitrary d and good SUSY only $E_0^- = 0$ will be exact. The associated normalized ground-state wave function reads

$$\phi_0^-(x) = \left[\frac{a}{\Gamma\left(\frac{1}{d+1}\right)} \left(\frac{d+1}{2a}\right)^{d/(d+1)} \right]^{1/2} \exp\left\{ -\frac{a}{d+1}|x|^{d+1} \right\}. \quad (6.63)$$

For $d = 1$ and good SUSY we arrive at the supersymmetric harmonic oscillator problem, where (6.62) provides the exact spectrum for all n. Formally, for good SUSY we may also include the case $d = 0$, if in this limit in the potential (6.60) the term $\pm ad|x|^{d-1}$ is interpreted as $\pm a\delta(x)$.[5] In this case only $n = 0$ is allowed.

Another case, where (6.62) becomes exact for all n and for good as well as broken SUSY, is the limit $d \to \infty$:

$$\lim_{d\to\infty} E_n^\pm = \frac{\hbar^2 \pi^2}{8m} \left(n + \frac{1}{2} \pm \frac{\Delta}{2} \right)^2. \quad (6.64)$$

That this is the exact spectrum can be seen by realizing that both, (6.60) and (6.61) become infinite square-well potentials in this limit:

$$\lim_{d\to\infty} V_\pm(x) = \begin{cases} 0 & \text{for } |x| < 1 \\ \infty & \text{for } |x| > 1 \end{cases}. \quad (6.65)$$

This limit has to be taken with some care. Actually, the Hilbert spaces $\mathcal{H}^\pm$ change in this limit, too: $L^2(\mathbb{R}) \to L^2([-1,1])$. In addition one has to specify boundary conditions at $x = \pm 1$ in order to have a well-defined problem. The

---

[5] Indeed, for $d = 0$ the SUSY potential (6.58) coincides with that for the δ-potential. See Example 4 on p. 31.

type of boundary conditions which has to be chosen is related to the requirement that the SUSY structure, that is, good or broken SUSY, is conserved in
the limit $d \to \infty$. It is obvious that for this one has to impose, for good SUSY,
Neumann conditions at $x = \pm 1$ for V_- and Dirichlet condition at $x = \pm 1$ for
V_+. Thus parity as well as SUSY remain to be good symmetries. For broken
SUSY (and, hence, broken parity) we have to choose for V_- Dirichlet conditions at $x = -1$ and Neumann conditions at $x = 1$, and vice versa for V_+. See
also our discussion of Example 3 and 4 in Sect. 3.5.3. That these boundary
conditions are indeed the one which are simulated by the finite-d system in
the limit $d \to \infty$ can also be seen by looking at the numerically calculated
energy eigenfunction of the Schrödinger equation [InJu94].

For finite d we have compared the eigenvalues (6.62) and those obtained
via the WKB formula (6.13) with the numerical exact eigenvalues of the
Schrödinger equation. Here we have chosen the parameter a to unity. Note
that the potentials (6.60) and (6.61) obey the scaling property

$$x \to \lambda x, \qquad a \to a/\lambda^{d+1}, \qquad E \to E/\lambda^2. \tag{6.66}$$

For the case of good SUSY we have considered the parameter values
$d = 2, 3, 5$. The results are given in Tables 6.2–6.4. The graphs for the corresponding partner potentials are shown in Fig. 6.3. Note that V_- is a double-
well potential with a barrier at $x = 0$. Hence, the WKB formula cannot be
applied for $n = 0$. This is indicated by the abbreviation N.A. in the tables.
These wells become smaller and deeper with increasing d and are located near
$x = \pm 1$.In the tables we also list in brackets the relative error

$$(E_{\text{exact}} - E_{\text{approx}})/E_{\text{exact}} \tag{6.67}$$

in percent. These relative errors are also shown in Figs. 6.5–6.7. Because of
the scaling property (6.66), the relative errors are independent of the coupling
parameter a. The case of good SUSY with odd d has also been studied by
Khare [Kha85]. See, however, the corrections made in [DuKhSu86].

In the case of broken SUSY we have made explicit calculations for the
parameter values $d = 2, 3, 4$, and 6. A graph of the corresponding potential
V_- is given in Fig. 6.4. Note that $V_+(x) = V_-(-x)$ in this case. The numerical
results are presented in Tables 6.5–6.8 and the relative errors are visualized
in Figs. 6.8–6.11. Note that, because of the symmetry between V_+ and V_-,
also the WKB approximation yields identical eigenvalues $E_n^- = E_n^+$ for both
potentials.

By inspecting the numerical results we see that the qc-SUSY approximation is in general not better than the usual WKB approximation. However, the
interesting observation we have made is, that the qc-SUSY estimate is always
above the WKB estimate:

$$E_{\text{qc-SUSY}} \geq E_{\text{WKB-}} \geq E_{\text{WKB+}}. \tag{6.68}$$

Furthermore, only for the case $d = 2$ and good SUSY we found one value $E_{\text{qc-SUSY}}$ (for $n = 2$), which is slightly below the exact eigenvalue. Note that the corresponding partner potentials are not differentiable at $x = 0$. See the solid line in Fig. 6.3. For this example we also find an unusual behavior of the relative error. That is, the absolute value of this error is not monotonically decreasing with increasing n. It rather shows an oscillatory behavior (see Fig. 6.4). Excluding such cases we arrive at the more interesting observation

$$E_{\text{qc-SUSY}} \geq E_{\text{exact}} \geq E_{\text{WKB+}} \tag{6.69}$$

for all continuously differentiable partner potentials $V_{\pm}$. In the case of broken SUSY we even found for such potentials that $E_{\text{WKB-}}$ is always below the exact value:

$$E_{\text{qc-SUSY}} \geq E_{\text{exact}} \geq E_{\text{WKB-}}. \tag{6.70}$$

We again emphasize that these three relations are not strict inequalities based on some mathematical proof, but rather express the observation made in our numerical investigations.

To support these relations we have also considered other systems, which do not belong to the above class of power SUSY potentials. Here we present results for exponential-type SUSY potentials

$$\Phi(x) = \frac{\hbar}{\sqrt{2m}} \sinh(x) \qquad \text{good SUSY},$$

$$\Phi(x) = \frac{\hbar}{\sqrt{2m}} \cosh(x) \qquad \text{broken SUSY}, \tag{6.71}$$

$$\Phi(x) = \frac{\hbar}{\sqrt{2m}} \exp\{x^2/2\} \qquad \text{broken SUSY}.$$

The numerical results displayed in Tables 6.9–6.11 and in Figs. 6.12–6.14 support our previously made observation.

In conclusion, we can say that the qc-SUSY approximation is as good as and in particular cases even better than the WKB formula. We have made the interesting observation that for some potentials $V_{\pm}$ the qc-SUSY approximation always overestimates the exact values. In the case of broken SUSY we, in addition, observe that the WKB approximation for some potentials always gives an underestimation. Because of this special feature, the approximation can be improved by taking an average of the results from these two approaches [InJu94]. Recent discussions which also include higher-order corrections to the WKB and CBC (good SUSY) formula [AdDuKhSu88, Var92] show that the above observation may only be true for the lowest-order as discussed here. Finally, let us note that a similar phenomenon, namely the ordering of energy levels of spherical symmetric potentials, has been observed within a WKB approximation [FeFuDe79] and a rigorous proof based on the algebra of the generalized creation and annihilation operator (3.3) has been given by Grosse et al. [GrMa84, BaGrMa84, Gro91]. For a derivation of lower and upper bounds to the ground state energy of a given Hamiltonian based on the factorization method see [Schm85].

Table 6.2. Comparison of the exact numerical eigenvalues with that obtained via the qc-SUSY and the WKB approximation for V_- (WKB$^-$) and V_+ (WKB$^+$) for $d = 2$ and good SUSY. Energies are given in units of $\hbar^2/2m$. The relative errors parenthesized are given in %

n	0	1	2	3	4	5	method
E_n^-	0.000	2.481	6.938	11.834	17.456	23.485	exact
E_n^-	0.000	2.753	6.937	11.912	17.481	23.538	qc-SUSY
		(+10.95)	(−0.14)	(+0.65)	(+0.14)	(+0.23)	
E_n^-	N.A.	2.462	6.821	11.817	17.409	23.475	WKB$^-$
		(+4.88)	(−1.55)	(−0.07)	(−0.27)	(−0.04)	
E_{n-1}^+	—	2.602	6.791	11.807	17.397	23.467	WKB$^+$
		(−0.76)	(−2.12)	(−0.23)	(−0.34)	(−0.08)	

Table 6.3. Same as Table 6.2 for $d = 3$ and good SUSY

n	0	1	2	3	4	5	method
E_n^-	0.000	2.302	7.490	13.889	21.451	30.021	exact
E_n^-	0.000	2.694	7.619	13.997	21.549	30.116	qc-SUSY
		(+17.04)	(+1.73)	(+0.78)	(+0.46)	(+0.32)	
E_n^-	N.A.	2.378	7.364	13.779	21.356	29.940	WKB$^-$
		(+3.34)	(−1.67)	(−0.80)	(−0.45)	(−0.27)	
E_{n-1}^+	—	2.123	7.372	13.728	21.323	29.917	WKB$^+$
		(−7.74)	(−7.74)	(−1.16)	(−0.60)	(−0.34)	

Table 6.4. Same as Table 6.2 for $d = 5$ and good SUSY

n	0	1	2	3	4	5	method
E_n^-	0.000	2.155	7.908	16.096	26.215	38.148	exact
E_n^-	0.000	2.625	8.333	16.378	26.454	38.372	qc-SUSY
		(+21.78)	(+5.37)	(+1.75)	(+0.91)	(+0.59)	
E_n^-	N.A.	2.010	7.743	15.829	25.939	37.885	WKB$^-$
		(−6.75)	(−2.09)	(−1.66)	(−1.05)	(−0.69)	
E_{n-1}^+	—	1.675	7.549	15.706	25.853	37.820	WKB$^+$
		(−22.26)	(−4.54)	(−2.42)	(−1.38)	(−0.86)	

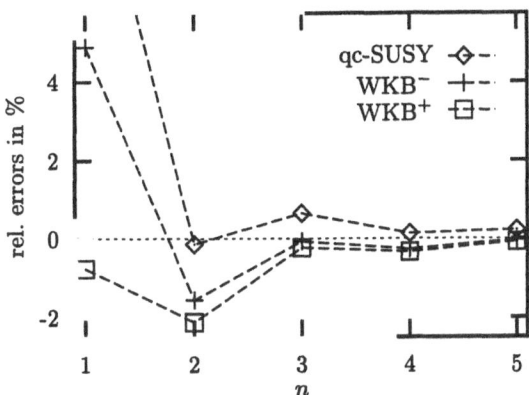

Fig. 6.5. Relative errors for the qc-SUSY and the WKB approximations for $d = 2$ and good SUSY.

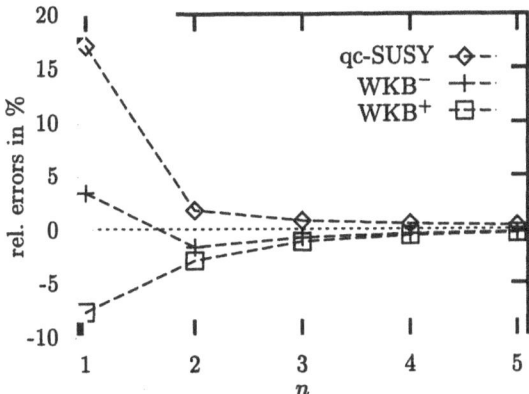

Fig. 6.6. Same as Fig. 6.5 for $d = 3$ and good SUSY.

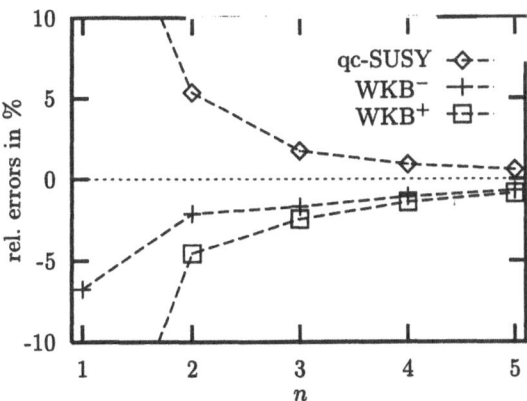

Fig. 6.7. Same as Fig. 6.5 for $d = 5$ and good SUSY.

Table 6.5. Same as Table 6.2 for $d = 2$ and broken SUSY

n	0	1	2	3	4	5	method
$E_n^{\pm}$	0.708	4.673	9.288	14.585	20.411	26.682	exact
$E_n^{\pm}$	1.093	4.727	9.341	14.630	20.453	26.728	qc-SUSY
	(+54.27)	(+1.16)	(+0.57)	(+0.31)	(+0.21)	(+0.17)	
$E_n^{\pm}$	0.658	4.571	9.233	14.544	20.381	26.665	WKB$^{\pm}$
	(−7.09)	(−2.20)	(−0.58)	(−0.28)	(−0.14)	(−0.07)	

Table 6.6. Same as Table 6.2 for $d = 3$ and broken SUSY

n	0	1	2	3	4	5	method
$E_n^{\pm}$	0.616	4.696	10.552	17.526	25.621	34.647	exact
$E_n^{\pm}$	0.952	4.949	10.648	17.638	25.636	34.745	qc-SUSY
	(+54.61)	(+5.38)	(+0.91)	(+0.64)	(+0.36)	(+0.28)	
$E_n^{\pm}$	0.357	4.598	10.380	17.413	25.516	34.566	WKB$^{\pm}$
	(−42.11)	(−2.09)	(−1.63)	(−0.65)	(−0.41)	(−0.23)	

Table 6.7. Same as Table 6.2 for $d = 4$ and broken SUSY

n	0	1	2	3	4	5	method
$E_n^{\pm}$	0.574	4.679	11.288	19.522	29.281	40.432	exact
$E_n^{\pm}$	0.875	5.076	11.494	19.692	29.438	40.583	qc-SUSY
	(+52.39)	(+8.47)	(+1.82)	(+0.87)	(+0.54)	(+0.38)	
$E_n^{\pm}$	0.207	4.528	11.043	19.297	29.081	40.254	WKB$^{\pm}$
	(−63.96)	(−3.24)	(−2.17)	(−1.15)	(−0.68)	(−0.44)	

Table 6.8. Same as Table 6.2 for $d = 6$ and broken SUSY

n	0	1	2	3	4	5	method
$E_n^{\pm}$	0.533	4.600	12.053	21.939	33.992	48.095	exact
$E_n^{\pm}$	0.793	5.217	12.524	22.297	34.304	48.389	qc-SUSY
	(+48.81)	(+13.42)	(+3.90)	(+1.63)	(+0.92)	(+0.61)	
$E_n^{\pm}$	0.083	4.314	11.686	21.520	33.575	47.697	WKB$^{\pm}$
	(−84.40)	(−6.21)	(−3.05)	(−1.91)	(−1.23)	(−0.83)	

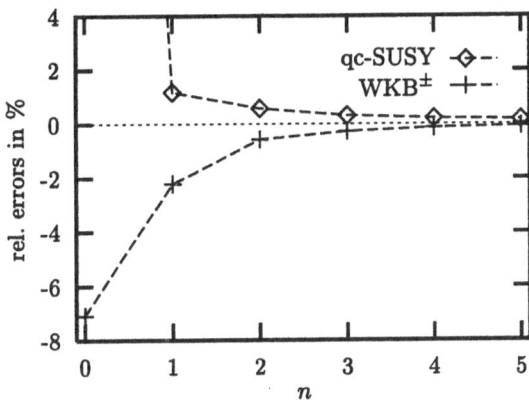

Fig. 6.8. Relative errors of the qc-SUSY and the WKB approximation for $d = 2$ and broken SUSY.

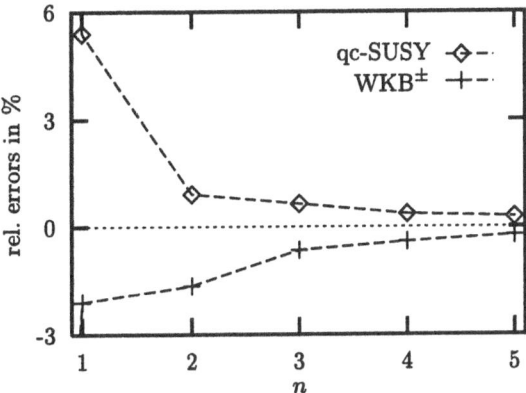

Fig. 6.9. Same as Fig. 6.8 for $d = 3$ and broken SUSY.

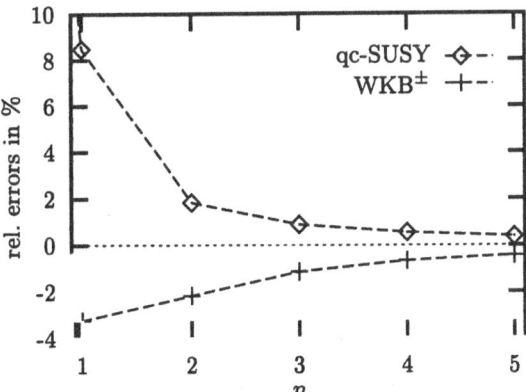

Fig. 6.10. Same as Fig. 6.8 for $d = 4$ and broken SUSY.

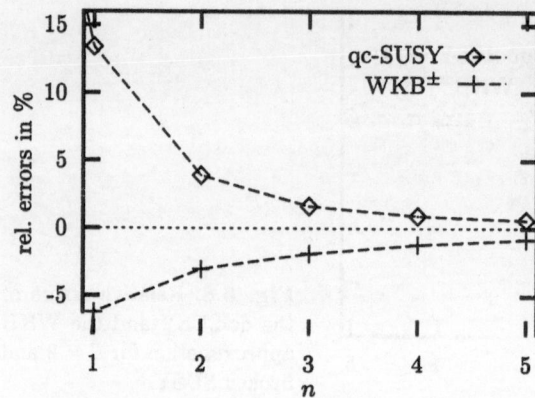

Fig. 6.11. Same as Fig. 6.8 for $d = 6$ and broken SUSY.

Table 6.9. Results for the good-SUSY potential $\Phi(x) = \sqrt{\hbar^2/2m}\,\sinh x$. Here the WKB approximation respects the exact relation $E_n^- = E_{n-1}^+$, $n = 1, 2, 3, \ldots$

n	0	1	2	3	4	5	method
$E_n^\pm$	0.000	3.245	7.229	11.811	16.919	22.506	exact
$E_n^\pm$	0.000	3.265	7.253	11.838	16.948	22.539	qc-SUSY
		(+0.61)	(+0.33)	(+0.22)	(+0.17)	(+0.15)	
E_n^-	−0.125	3.155	7.152	11.742	16.857	22.450	WKB$^-$
		(−2.77)	(−1.07)	(−0.59)	(−0.37)	(−0.25)	

Table 6.10. Results for the broken-SUSY potential $\Phi(x) = \sqrt{\hbar^2/2m}\,\cosh x$

n	0	1	2	3	4	5	method
$E_n^\pm$	3.438	7.116	11.424	16.284	21.642	27.458	exact
$E_n^\pm$	3.530	7.178	11.476	16.330	21.686	27.503	qc-SUSY
	(+2.68)	(+0.88)	(+0.45)	(+0.28)	(+0.20)	(+0.16)	
$E_n^\pm$	3.338	7.033	11.351	16.218	21.581	27.404	WKB$^\pm$
	(−2.92)	(−1.16)	(−0.64)	(−0.41)	(−0.28)	(−0.20)	

Table 6.11. Results for the broken-SUSY potential $\Phi(x) = \sqrt{\hbar^2/2m}\,\exp\{x^2/2\}$

n	0	1	2	3	4	5	method
$E_n^\pm$	3.533	7.678	12.959	19.263	26.521	34.682	exact
$E_n^\pm$	3.593	7.742	13.025	19.333	26.595	34.762	qc-SUSY
	(+1.71)	(+0.83)	(+0.51)	(+0.36)	(+0.28)	(+0.23)	
$E_n^\pm$	3.337	7.491	12.775	19.083	26.345	34.512	WKB$^\pm$
	(−5.54)	(−2.44)	(−1.42)	(−0.94)	(−0.66)	(−0.49)	

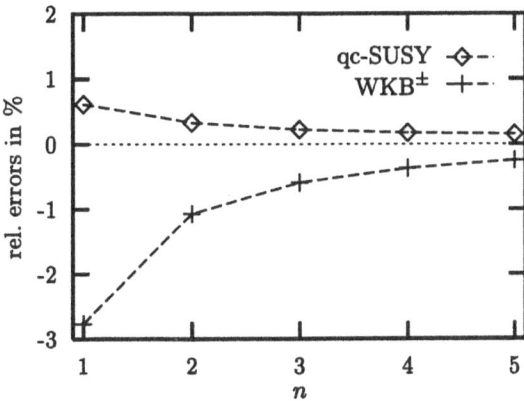

Fig. 6.12. Relative errors for the good-SUSY potential $\Phi(x) = \sqrt{\hbar^2/2m}\ \sinh x$.

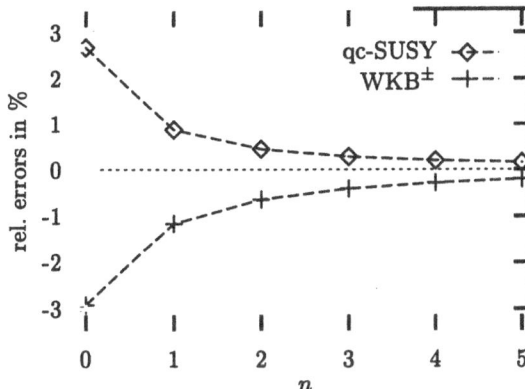

Fig. 6.13. Relative errors for the broken-SUSY potential $\Phi(x) = \sqrt{\hbar^2/2m}\ \cosh x$.

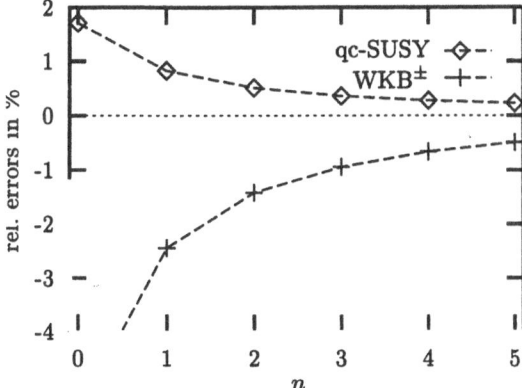

Fig. 6.14. Relative errors for the broken-SUSY potential $\Phi(x) = \sqrt{\hbar^2/2m}\ \exp\{x^2/2\}$.

7. Supersymmetry
in Classical Stochastic Dynamics

In 1979 Parisi and Sourlas [PaSo79] pointed out that there is a hidden supersymmetry in classical stochastic differential equations. In fact, it is possible to reformulate some supersymmetric models of field theory in terms of classical stochastic equations [PaSo82]. The existence or non-existence of a stationary solution of a classical stochastic system is related to good and broken SUSY in the corresponding field theory, respectively [FeTs82]. In the case of one cartesian degree of freedom, whose averaged dynamics is characterized by a one-dimensional Fokker–Planck equation, it can be shown [Kam81, PaSo82, CoFr83, BeBr84] that this equation may be put into the form of a supersymmetric Schrödinger equation with imaginary time. The corresponding Hamiltonian can be identified with that of Witten's model. This analogy explains the fact that the Fokker–Planck operator for a given drift potential and diffusion constant is essential iso-spectral to the Fokker–Planck operator with the inverted drift potential and the same diffusion constant [BeBr84, Ris89]. The breaking of SUSY in the Langevin dynamics of non-potential systems is also related to the occurrence of corrections to the linear fluctuation-dissipation theorem [Tri90].

In this chapter, we will review the relations between the classical stochastic dynamics of one cartesian degree of freedom and supersymmetric quantum mechanics. We will show that SUSY may be utilized to derive decay rates for bistable and metastable Fokker–Planck equations. For applications to field theories of statistical and condensed matter physics see, for example, [Sou85, ZiJu93].

7.1 Langevin and Fokker–Planck Equation

The dynamics of many complex systems in physics, chemistry and biology can be described phenomenologically by the so-called *Langevin equation* [Lan08, Arn73, Kam81, Gar90]

$$\dot{\eta} = F(\eta) + \xi(t). \tag{7.1}$$

Here $\eta \in \mathbb{R}$ denotes a macroscopic degree of freedom, the time-evolution of which we are interested in. The quantity $\xi \in \mathbb{R}$ is assumed to be a random function of time. The stochastic differential equation (7.1) first appeared in the work of Langevin [Lan08] who studied a simplified approach of Einstein's [Ein05, Ein06] and Smoluchowski's [Smo06] description of Brownian motion. In this model η denotes the momentum (or for highly overdamped motion the position) of the Brownian particle, F stands for an external deterministic force and ξ is a random function characterizing the fluctuations of the medium in which the Brownian particle is immersed. Properties of this random function, which is usually called *noise*, are characterized by expectation values. Without loss of generality, one may assume

$$\langle \xi(t) \rangle = 0, \qquad t \geq 0, \tag{7.2}$$

because a non-vanishing expectation value can be absorbed in the deterministic force F. The Langevin equation (7.1), as it stands, may characterize physical systems with noise ξ correlated on a given time scale τ_c as, for example [Gar90], like

$$\langle \xi(t)\xi(t') \rangle = \frac{D}{2\tau_c} \exp\{-|t - t'|/\tau_c\}, \qquad D > 0. \tag{7.3}$$

Such a noise is called *colored noise* and is a realistic description of various physical systems. For many systems, however, it is justified to consider the idealization of so-called δ-correlated or *white noise* by taking the limit $\tau_c \searrow 0$:

$$\langle \xi(t)\xi(t') \rangle = D\delta(t - t'). \tag{7.4}$$

To make the problem definite, in addition to (7.4), we will assume that this white noise has a Gaussian distribution, which is uniquely characterized by (7.2) and (7.4):

$$\langle \cdot \rangle := \int \mathcal{D}\xi \exp\left\{ -\frac{1}{2D} \int\limits_0^\infty d\tau\, \xi^2(\tau) \right\} (\cdot). \tag{7.5}$$

Usually, one is not interested in a particular solution of (7.1) for a given realization ξ. Of interest are, however, properties of typical realizations. These properties may be deduced from the probability density for arriving at $x \in \mathbb{R}$ in a given time t if the particle initially started in $x_0 \in \mathbb{R}$:

$$m_t(x, x_0) := \langle \delta(\eta(t) - x) \rangle, \qquad x_0 := \eta(0). \tag{7.6}$$

This transition-probability density can be shown [Lesch81, Kam81] to be determined by the so-called *Fokker–Planck equation* [Fok14, Pla17]

$$\frac{\partial}{\partial t} m_t(x, x_0) = \frac{D}{2} \frac{\partial^2}{\partial x^2} m_t(x, x_0) + \frac{\partial}{\partial x} F(x) m_t(x, x_0),$$

$$m_0(x, x_0) = \langle \delta(x - x_0) \rangle. \tag{7.7}$$

This is a diffusion equation with diffusion constant $D/2$ and an additional drift coefficient F.

In the following we will show that the Fokker–Planck equation can be put into the form of an imaginary-time Schrödinger equation with supersymmetry of the type of Witten's model. The drift coefficient F will turn out to play the role of the SUSY potential Φ. As a consequence the diffusion problem with drift $F = \Phi$ will be related via SUSY to the same problem with inverted drift $F = -\Phi$.

7.2 Supersymmetry of the Fokker–Planck Equation

In this section we will study the Fokker–Planck equation for two drift coefficients which differ by an overall sign, $F_\pm := \pm\Phi$. Being a deterministic force in one dimension, we may introduce the drift potential $U'_\pm = -F_\pm$. In the present case we have the pair

$$U_\pm(x) := \mp \int_0^x dz\, \Phi(z) = -U_\mp(x). \tag{7.8}$$

Typical drift potentials $U_\pm$ and their related drift coefficients are shown in Fig. 7.1. Here we adopt the convention that the stable drift potential, if there is any, is given by U_-. This convention will turn out to be equivalent to our convention (3.19) made for the Witten model. The corresponding Fokker–Planck equation reads

$$\frac{\partial}{\partial t} m_t^\pm(x, x_0) = \frac{D}{2} \frac{\partial^2}{\partial x^2} m_t^\pm(x, x_0) \pm \frac{\partial}{\partial x} \Phi(x) m_t^\pm(x, x_0), \tag{7.9}$$

where we have introduced an additional superscript in the transition-probability density in order to discriminate the solution for the drift potential U_- from the one for U_+.

Making the ansatz [Kam81]

$$m_t^\pm(x, x_0) =: \exp\left\{-\frac{1}{D}[U_\pm(x) - U_\pm(x_0)]\right\} K_\pm(x, t),$$

$$K_\pm(x, 0) = \langle\delta(x - x_0)\rangle, \tag{7.10}$$

one arrives at the imaginary-time Schrödinger equation

$$-D\frac{\partial}{\partial t} K_\pm(x, t) = H_\pm K_\pm(x, t), \tag{7.11}$$

where

$$H_\pm := -\frac{D^2}{2} \frac{\partial^2}{\partial x^2} + \frac{1}{2} \Phi^2(x) \pm \frac{D}{2} \Phi'(x) \tag{7.12}$$

appear to be the partner Hamiltonians of Witten's model for unit mass upon the substitution

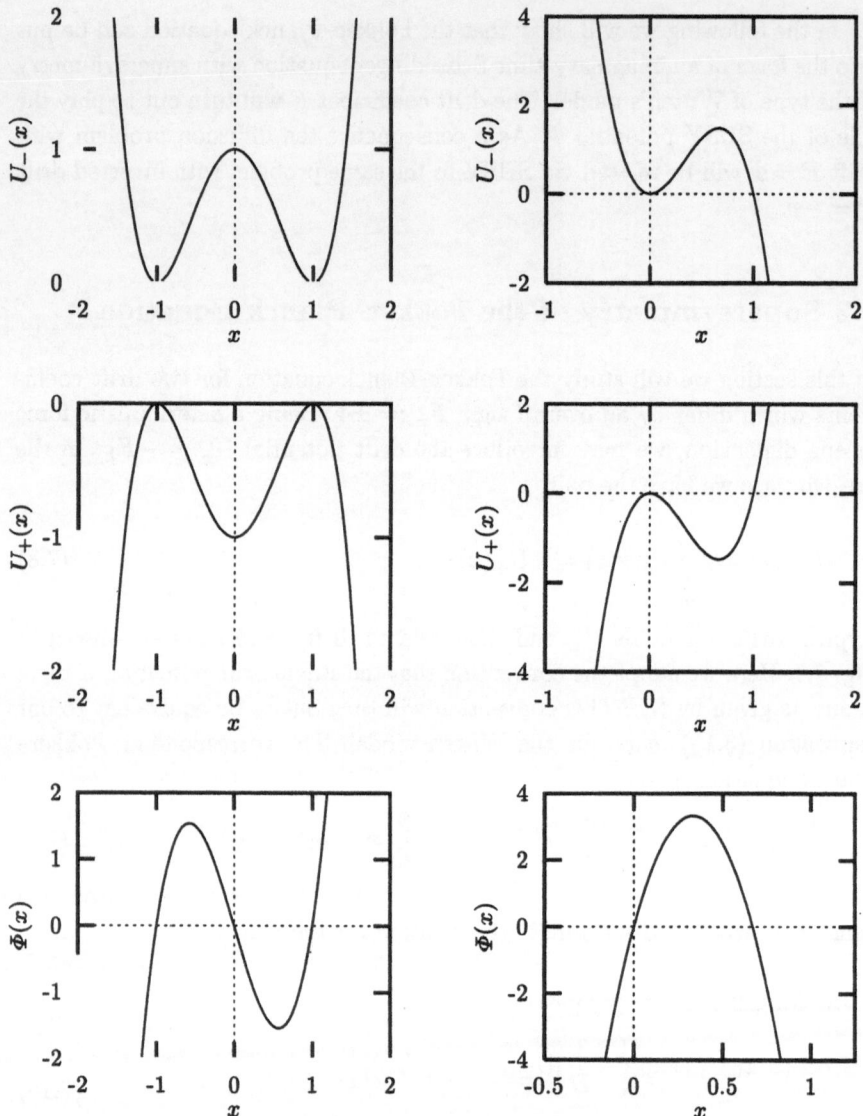

Fig. 7.1. Typical drift potentials U_- and inverted drift potentials $U_+ = -U_-$ for drift coefficients Φ with good SUSY (*left row*) and broken SUSY (*right row*).

$$\Phi \to \sqrt{2}\,\Phi, \qquad D \to \hbar. \tag{7.13}$$

Clearly, the drift coefficient Φ plays the role of the SUSY potential and, hence, the drift potential U_- may be identified with the superpotential (3.12). With the initial condition (7.10) the solution of (7.11) is given by the heat kernel

$$K_\pm(x,t) = \langle x| \exp\{-tH_\pm/D\}|x_0\rangle. \tag{7.14}$$

Denoting the eigenvalues of $H_\pm$ by $\lambda_n^\pm$ (we will assume a purely discrete spectrum for simplicity) and the associated eigenfunctions by $\phi_n^\pm$, that is,

$$H_\pm \phi_n^\pm(x) = \lambda_n^\pm \phi_n^\pm(x), \qquad n = 0, 1, 2, \ldots, \tag{7.15}$$

we arrive at the spectral representation [Kam81]

$$
\begin{aligned}
m_t^\pm(x, x_0) = \exp &\left\{ -\frac{1}{D}\left[U_\pm(x) - U_\pm(x_0) \right] \right\} \\
&\times \sum_{n=0}^\infty \exp\left\{ -\frac{1}{D} t\lambda_n^\pm \right\} \phi_n^\pm(x)\phi_n^\pm(x_0).
\end{aligned}
\tag{7.16}
$$

Because of SUSY we know that H_- and H_+ are essential iso-spectral. That is, the strictly positive decay rates for the drift potential U_- and the inverted one $U_+ = -U_-$ are identical [Ris89]. This fact is even true for colored noise, where it is also related in some sense to an underlying SUSY [LeMaRi87].

In addition, in the case of a good SUSY, we have a vanishing eigenvalue $\lambda_0^- = 0$. As a consequence, good SUSY implies the existence of a stationary non-trivial distribution

$$P(x) := \lim_{t\to\infty} m_t^-(x, x_0), \tag{7.17}$$

which, due to our convention, belongs to U_-.

7.3 Supersymmetry of the Langevin Equation

The Langevin equation associated with the Fokker–Planck equation (7.9) of the previous section reads

$$\dot{\eta} = \pm \Phi(\eta) + \xi^\pm(t), \tag{7.18}$$

where, as in Sect. 7.1, ξ^+ and ξ^- denote independent Gaussian white-noise variables with the same diffusion constant $D = 1$:

$$\langle \xi^+(t)\xi^+(t')\rangle_+ = \langle \xi^-(t)\xi^-(t')\rangle_- = \delta(t - t'),$$

$$\langle \cdot \rangle_\pm := \int \mathcal{D}\xi^\pm \exp\left\{ -\frac{1}{2}\int_0^\infty d\tau\, [\xi^\pm(\tau)]^2 \right\}(\cdot). \tag{7.19}$$

It is interesting to note that $\xi^\pm = \dot{\eta} \pm \Phi(\eta)$ are the Euclidean versions of the classical canonical momenta introduced in (4.26). In other words, they may be considered as the canonical momenta of the Euclidean version of the quasi-classical Lagrangian (4.25):

$$\tilde{L}^\pm_{\text{qc, eucl.}}(\dot{\eta}, \eta) := \frac{1}{2}\left(\dot{\eta} \mp \Phi(\eta)\right)^2. \tag{7.20}$$

Let us now consider the transition-probability density associated with the Langevin equation (7.18):

$$m_t^\pm(x, x_0) = \langle\delta(\eta(t) - x)\rangle_\pm, \qquad x_0 = \eta(0). \tag{7.21}$$

Using the definition (7.19) we find the following path-integral representation

$$m_t^\pm(x, x_0) = \int\limits_{\eta(0)=x_0} \mathcal{D}\xi^\pm \exp\left\{-\frac{1}{2}\int\limits_0^\infty \mathrm{d}\tau\,[\xi^\pm(\tau)]^2\right\}\delta(\eta(t) - x). \tag{7.22}$$

This Gaussian path integral may be transformed into a Wiener-type path integral upon the transformation $\xi^\pm \to \eta$, which is called Nicolai map [CeGi83, KiSaSk85, EzKl85, GrRo85, SiSt86]:

$$\xi^\pm(t) = \dot{\eta}(t) \mp \Phi(\eta(t)). \tag{7.23}$$

Noting that

$$\frac{\delta\xi^\pm(t)}{\delta\eta(t')} = \left(\frac{\partial}{\partial t} \mp \Phi'(\eta(t))\right)\delta(t - t') \tag{7.24}$$

we find

$$
\begin{aligned}
m_t^\pm(x, x_0) &= \int\limits_{\eta(0)=x_0} \mathcal{D}\eta \exp\left\{-\frac{1}{2}\int\limits_0^\infty \mathrm{d}\tau\,[\dot{\eta}(\tau) \mp \Phi(\eta(\tau))]^2\right\} \\
&\quad \times \delta(\eta(t) - x)\det\left[\left(\frac{\partial}{\partial t} \mp \Phi'(\eta(t))\right)\delta(t - t')\right] \\
&= \int\limits_{\eta(0)=x_0}^{\eta(t)=x} \mathcal{D}\eta \exp\left\{-\int\limits_0^t \mathrm{d}\tau\,\tilde{L}^\pm_{\text{qc, eucl.}}(\dot{\eta}(\tau), \eta(\tau))\right\} \\
&\quad \times \det\left[\left(\frac{\partial}{\partial t} \mp \Phi'(\eta(t))\right)\delta(t - t')\right].
\end{aligned}
\tag{7.25}
$$

Here several remarks are in order. The functional determinant appearing in (7.25) can be given a definite meaning if the associated Nicolai map (7.23) is properly interpreted as a change from the stochastic Gaussian process in $\xi^\pm$ to a Wiener process in η. Ezawa and Klauder [EzKl85] have considered a Stratonovich- (Str) and an Itô-related interpretation of (7.23) leading to

$$\det \left[\frac{\delta \xi^{\pm}(t)}{\delta \eta(t')} \right]_{\text{Str}} = \text{const.} \times \exp \left\{ \mp \frac{1}{2} \int_0^t d\tau \, \Phi'(\eta(\tau)) \right\},$$

$$\det \left[\frac{\delta \xi^{\pm}(t)}{\delta \eta(t')} \right]_{\text{Itô}} = \text{const.} \times 1, \tag{7.26}$$

where the (infinite) constant may be absorbed by the normalization factor. Similar, the Euclidean action appearing in the exponent of (7.25) has to be interpreted in a consistent way:

$$\int_0^t d\tau \tilde{L}^{\pm}_{\text{qc, eucl.}}(\dot{\eta}(\tau), \eta(\tau)) = \frac{1}{2} \int_0^t d\tau \, [\dot{\eta}^2(\tau) + \Phi^2(\eta(\tau))]$$

$$\mp \int_0^t d\tau \, \dot{\eta}(\tau) \Phi(\eta(\tau)). \tag{7.27}$$

It is the last integral of this relation which requires a proper interpretation, because of the stochastic nature of $\eta(\tau)$. For the Stratonovich, respectively, the Itô interpretation we have [EzKl85]

$$\pm \int_0^t d\tau \, \dot{\eta}(\tau) \, \Phi(\eta(\tau)) \bigg|_{\text{Str}} = \pm \int_{x_0}^x dz \, \Phi(z) = - \left[U_{\pm}(x) - U_{\pm}(x_0) \right],$$

$$\pm \int_0^t d\tau \, \dot{\eta}(\tau) \, \Phi(\eta(\tau)) \bigg|_{\text{Itô}} = - \left[U_{\pm}(x) - U_{\pm}(x_0) \right] \mp \frac{1}{2} \int_0^t d\tau \, \Phi'(\eta(\tau)), \tag{7.28}$$

where $U_{\pm}$ is defined in (7.8). Both interpretations lead, of course, to the same Wiener-type path-integral expression

$$m_t^{\pm}(x, x_0) = \exp \left\{ - \left[U_{\pm}(x) - U_{\pm}(x_0) \right] \right\}$$

$$\times \int_{\eta(0)=x_0}^{\eta(\tau)=x} \mathcal{D}\eta \exp \left\{ - \frac{1}{2} \int_0^t d\tau \, [\dot{\eta}^2(\tau) + \Phi^2(\eta(\tau)) \pm \Phi'(\eta(\tau))] \right\} \tag{7.29}$$

$$= \exp \left\{ - \left[U_{\pm}(x) - U_{\pm}(x_0) \right] \right\} \langle x | e^{-tH_{\pm}} | x_0 \rangle,$$

where $H_{\pm}$ is given in (7.12). This result coincides, as expected, with (7.10–7.14).

Let us note, that the functional determinant in (7.25) may also be expressed in terms of a fermionic path integral

$$\det \left[\left(\frac{\partial}{\partial t} \mp \Phi'(\eta(t)) \right) \delta(t - t') \right]_{\text{Str}}$$

$$= \int \mathcal{D}\psi \int \mathcal{D}\bar{\psi} \exp \left\{ \int_0^t d\tau \, \bar{\psi}(\tau) \left(\frac{\partial}{\partial \tau} \mp \Phi'(\eta(\tau)) \right) \psi(\tau) \right\}. \tag{7.30}$$

Here we have taken the Stratonovich-related interpretation for the determinant.[1] The fermionic path integral, as it stands, is not well defined. To give it a meaning Ezawa and Klauder [EzKl85] used a time-lattice definition $\epsilon := t/N$, $\psi_j := \psi(j\epsilon)$, and showed that the right-hand-side of (7.30) may be given by[2]

$$\lim_{N\to\infty} \int \prod_{j=0}^{N} (\mathrm{d}\psi_j \mathrm{d}\bar\psi_j)$$

$$\times \exp\left\{ \sum_{j=0}^{N} \left[\bar\psi_j(\psi_j - \psi_{j-1}) \mp \frac{\epsilon}{2}\Phi'(\eta(\epsilon j))\bar\psi_j(\psi_j + \psi_{j-1}) \right] \right\}, \tag{7.31}$$

where $\psi_{-1} := 0$, which they called half-Dirichlet boundary condition. Having this in mind we may represent the transition-probability density as follows

$$m_t^\pm(x, x_0) = \int_{\eta(0)=x_0}^{\eta(t)=x} \mathcal{D}\eta \int \mathcal{D}\psi \int \mathcal{D}\bar\psi \exp\left\{ -\int_0^t \mathrm{d}\tau\, L_0^\pm \right\}, \tag{7.32}$$

where

$$L_0^\pm := \frac{1}{2}\dot\eta^2 + \frac{1}{2}\Phi^2(\eta) - \bar\psi\dot\psi \pm \Phi'(\eta)\bar\psi\psi. \tag{7.33}$$

The Lagrangian $L_0^\pm$ is, up to a total time derivative, invariant under the SUSY transformations

$$\delta\eta = \bar\epsilon\psi - \bar\psi\epsilon, \qquad \delta\psi = \epsilon(\dot\eta \pm \Phi(\eta)), \qquad \delta\bar\psi = (\dot\eta \mp \Phi(\eta))\bar\epsilon. \tag{7.34}$$

There exists also another time-lattice definition for the right-hand-side of (7.30) which reads (cf. eq. (A.7) of [EzKl85])

$$\lim_{N\to\infty} \int \prod_{j=0}^{N} (\mathrm{d}\psi_j \mathrm{d}\bar\psi_j)$$

$$\times \psi_N \exp\left\{ \sum_{j=0}^{N} \left[\bar\psi_j(\psi_j - \psi_{j-1}) \pm \frac{\epsilon}{2}\Phi'(\eta(\epsilon j))\bar\psi_j(\psi_j + \psi_{j-1}) \right] \right\} \bar\psi_0, \tag{7.35}$$

where again the half-Dirichlet boundary condition $\psi_{-1} := 0$ is used. Here, in essence, the overall sign of the SUSY potential Φ has been changed and the exponential is now sandwiched between $\psi_N = \psi(t)$ and $\bar\psi_0 = \bar\psi(0)$. With this definition we can put the transition-probability density into the form

---

[1] Because of (7.26) the Itô-related interpretation will only lead to a trivial fermion path integral. There would be no coupling between the bosonic variable η and the fermionic variables ψ, $\bar\psi$.

[2] See eq. (3.19) in [EzKl85]. Note, however, that our convention (4.63) for the Grassmann integration rule differs from that of Ezawa and Klauder.

$$m_t^{\pm}(x, x_0) = \int\limits_{\eta(0)=x_0}^{\eta(t)=x} \mathcal{D}\eta \int \mathcal{D}\psi \int \mathcal{D}\bar{\psi}\, \psi(t) \exp\left\{ -\int_0^t d\tau\, L_1^{\pm} \right\} \bar{\psi}(0), \qquad (7.36)$$

where

$$L_1^{\pm} := \frac{1}{2}\dot{\eta}^2 + \frac{1}{2}\Phi^2(\eta) - \bar{\psi}\dot{\psi} \mp \Phi'(\eta)\bar{\psi}\psi. \qquad (7.37)$$

Again, because of $L_1^{\pm} = L_0^{\mp}$, this Lagrangian is invariant under the SUSY transformation (7.34).

It should be noted, that these two representations of the formal fermionic path integral (7.30) are related to the invariance of Witten's model under a change of an overall sign of the SUSY potential. See our discussion of Sect. 3.3.2. This two-fold SUSY has been interpreted by Gozzi [Goz84] as macroscopic manifestation of Onsagers principle of microscopic reversibility. Indeed, the backward Fokker-Planck equation can be obtained from (7.9) by taking its adjoint on the right-hand-side. This, however, is equivalent to a change of the overall sign of the SUSY potential. In the following we will study some implications of this symmetry between the forward and the backward process.

7.4 Implications of Supersymmetry

7.4.1 Good SUSY

First we will consider the good SUSY case. Typical drift and SUSY potentials are shown in the left row of Fig. 7.1. From our discussion of the Witten model we know that the decay rates $\lambda_n^{\pm}$ in the potentials $U_{\pm}$ are related by

$$\lambda_n := \lambda_n^- = \lambda_{n-1}^+ > 0, \qquad n = 1, 2, 3, \ldots,$$
$$\lambda_0^- = 0, \qquad\qquad\qquad\qquad\qquad\qquad (7.38)$$

and the ground-state wave function of H_- is given by

$$\phi_0^-(x) = C \exp\left\{ -\frac{1}{D}\int_0^x dz\, \Phi(z) \right\} = C \exp\left\{ -\frac{1}{D}U_-(x) \right\}, \qquad (7.39)$$

where C denotes the normalization constant. The transition-probability densities read

$$m_t^-(x, x_0) = |\phi_0^-(x)|^2 + \frac{\phi_0^-(x)}{\phi_0^-(x_0)} \sum_{n=1}^{\infty} e^{-\lambda_n t/D}\phi_n^-(x)\phi_n^-(x_0),$$
$$m_t^+(x, x_0) = \frac{\phi_0^-(x_0)}{\phi_0^-(x)} \sum_{n=1}^{\infty} e^{-\lambda_n t/D}\phi_{n-1}^+(x)\phi_{n-1}^+(x_0). \qquad (7.40)$$

As already mentioned, because of good SUSY, there exists a stationary distribution for U_-, which is given by the SUSY ground state

$$P(x) = |\phi_0^-(x)|^2 = C^2 \exp\left\{-\frac{2}{D}U_-(x)\right\}. \tag{7.41}$$

Due to the SUSY transformation (3.9) we can also relate the normalized decay modes $\phi_n^\pm$ of U_- with that of U_+ and vice versa:

$$\phi_{n-1}^+(x) = \frac{1}{\sqrt{2\lambda_n}}\left(D\frac{\partial}{\partial x} + \Phi(x)\right)\phi_n^-(x),$$

$$\phi_n^-(x) = \frac{1}{\sqrt{2\lambda_n}}\left(D\frac{\partial}{\partial x} - \Phi(x)\right)\phi_{n-1}^+(x), \qquad n = 1,2,3,\ldots. \tag{7.42}$$

SUSY in not only useful for obtaining the above connections between the diffusion in a given drift potential and the diffusion in the inverted potential. It also proofs to be useful for explicit calculations. For example, for all shape-invariant SUSY potentials listed in Table 5.1 one can easily find exact solutions of the associated Fokker–Planck equation [HoZh82, Eng88, Jau88]. Approximate results can also be obtained, for example, via a variational approach. Here, in essence, one obtains from a variational ansatz the ground-state properties of H_+, that is, λ_1 and ϕ_0^+. Then via (7.38) and (7.42) one finds the timescale $1/\lambda_1$ and mode ϕ_1^- with which the system U_- relaxes into the stationary distribution (7.41). This process can in principle be continued to obtain a hierarchy of diffusion potentials in analogy to our discussion at the end of Sect. 5.1.

Here we limit ourself to an approximate derivation of the smallest non-vanishing eigenvalue λ_1 based on a SUSY method originally devised for tunneling problems [KeKoSu88, Wit92]. We want to find the ground-state properties of the Hamiltonian H_+ of the pair of partner Hamiltonians

$$H_\pm = -\frac{D}{2}\frac{\partial^2}{\partial x^2} + V_\pm(x) \tag{7.43}$$

with partner potentials

$$V_\pm(x) := \frac{1}{2}\Phi^2(x) \pm \frac{D}{2}\Phi'(x). \tag{7.44}$$

See left row of Fig. 7.2 for a plot of V_-, V_+ and Φ^2, which are associated with the drift potential given in the left row of Fig. 7.1. For simplicity, we will assume that Φ is an odd function, which implies that $U_\pm$ and $V_\pm$ are even functions. We further assume that U_- is a bistable potential as shown in Fig. 7.1 left row (for the corresponding partner potentials see Fig. 7.2 left row) with a high barrier between the two minima. The last assumption implies that the eigenvalue $\lambda_1 > 0$ will be very close to zero. Here we note that a

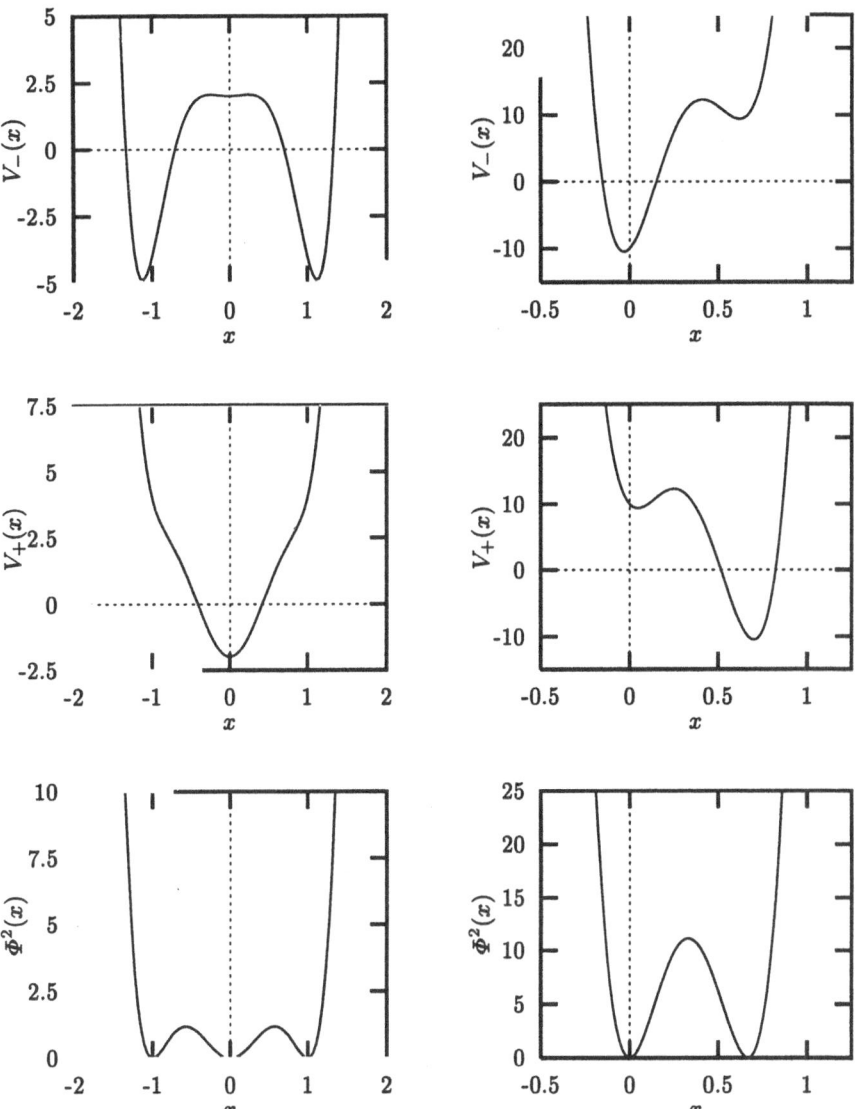

Fig. 7.2. The partner potentials $V_\pm$ and Φ^2 for the drift potentials shown in Fig. 7.1. Again the left row corresponds to the good-SUSY and the right to the broken-SUSY case. The diffusion constant D has been set to unity.

non-normalizable eigenfunction of H_+ with zero eigenvalue is given by (cf. eq. (3.14))

$$\frac{1}{\phi_0^-(x)} \propto \exp\{U_-(x)/D\} = \exp\{-U_+(x)/D\}. \tag{7.45}$$

Obviously, for $x \to \pm\infty$ this function diverges. However, for small values of x it may be a good approximation to the exact ground state ϕ_0^+ of H_+. In fact, regularizing (7.45) for large x values one can use it as a trial function for a variational approach [BeBr84, MaSoZa88].

Here, however, we choose a perturbative approach. Let us consider the following nodeless and normalizable wave function [KeKoSu88, Wit92]

$$\phi(x) := \frac{C}{\phi_0^-(|x|)} \int\limits_{|x|}^{+\infty} dz \, \left(\phi_0^-(z)\right)^2. \tag{7.46}$$

For small x values this wave function is similar to (7.45) and therefore is a good approximation to ϕ_0^+. It is, however, not differentiable at $x = 0$. In fact, it is easily verified that ϕ is the exact zero-energy ground-state wave function for the Hamiltonian

$$H := H_+ - 2D^2 \left(\phi_0^-(0)\right)^2 \delta(x). \tag{7.47}$$

The Hamiltonian H_+ may be viewed as Hamiltonian H with additional δ-like perturbation. Knowing the ground-state properties of H one may obtain those of H_+ via a perturbation expansion. In a first-order perturbation expansion the ground-state eigenvalue of H_+ is given as

$$\lambda_1 \simeq \int\limits_{\mathbb{R}} dx \, (H_+ - H)\phi^2(x) = 2D^2 \left(\phi_0^-(0)\right)^2 \phi^2(0) = \frac{D^2}{2} C^2. \tag{7.48}$$

Hence, the eigenvalue λ_1 is essentially given by the normalization constant C introduced in (7.39). Using the Laplace method for small D in the calculation of C one arrives [Wit92] at the result of Kramers [Kra40]

$$\lambda_1 \simeq \frac{D}{\pi} \sqrt{U_-''(x_{min})|U_-''(x_{max})|}$$
$$\times \exp\left\{-\frac{2}{D}\left[U_-(x_{max}) - U_-(x_{min})\right]\right\}. \tag{7.49}$$

Here x_{min} is the position of the (right) minimum of U_- ($x_{min} = 1$ in Fig. 7.1 left row) and x_{max} is the position of the barrier of U_- ($x_{max} = 0$ in Fig. 7.1 left row). Despite the fact that this derivation has been given for a symmetric drift potential it can be used for asymmetric potentials, too [Ris89].

Higher eigenvalues λ_n may be estimated via the WKB approximation $[V_\pm(q_L) = \lambda = V_\pm(q_R)]$

$$\int\limits_{q_L}^{q_R} dx \sqrt{2(\lambda - V_\pm(x))} = D\pi \left(n + \tfrac{1}{2}\right) \tag{7.50}$$

or via the qc-SUSY approximation $[\Phi^2(x_L) = 2\lambda = \Phi^2(x_R)]$

$$\int\limits_{x_L}^{x_R} dx \sqrt{2\lambda - \Phi^2(x)} = D\pi n. \tag{7.51}$$

These approximations may even be applied for an approximation to λ_1 if $D \sim U(x_{max}) - U(x_{min})$, where the Kramers result is not applicable. Our discussion of Sect. 6.4.2 in addition shows that a combination of both approximations may yield lower and upper bounds to the exact decay rates.

7.4.2 Broken SUSY

The discussion for the broken-SUSY case is similar to the previous one. Here the decay rates for $U_\pm$ (see Fig. 7.1 right row) are identical, that is,

$$\lambda_n := \lambda_n^- = \lambda_n^+ > 0, \qquad n = 0, 1, 2, \ldots. \tag{7.52}$$

The corresponding transition-probability densities are

$$\boxed{\begin{aligned} m_t^\pm(x, x_0) &= \exp\left\{ \pm \frac{1}{D}[U_-(x) - U_-(x_0)] \right\} \\ &\quad \times \sum_{n=0}^{\infty} e^{-\lambda_n t/D} \phi_n^\pm(x)\phi_n^\pm(x_0), \end{aligned}} \tag{7.53}$$

where the decay modes are related by

$$\phi_n^\pm(x) = \frac{1}{\sqrt{2\lambda_n}}\left(D\frac{\partial}{\partial x} \pm \Phi(x)\right)\phi_n^\mp(x). \tag{7.54}$$

Note that because of broken SUSY ($\lambda_0 > 0$) there exists no stationary distribution.

As for the good-SUSY case there are shape-invariant SUSY potentials with broken SUSY. Hence, it is also possible to study the unstable or metastable Fokker–Planck equation exactly. As an interesting example we mention

$$\Phi(x) := \Phi_0 \tanh x + a, \qquad \Phi_0 > 0, \; a > 0,$$

$$U_-(x) = \Phi_0 \ln(\cosh x) + ax. \tag{7.55}$$

For $a < \Phi_0$ SUSY is good (see Table 5.1). However, for $a \geq \Phi_0$ SUSY will be broken. The corresponding drift potential U_- for various values of a is shown in Fig. 7.3. Of particular interest is the broken case $a = \Phi_0$, which describes

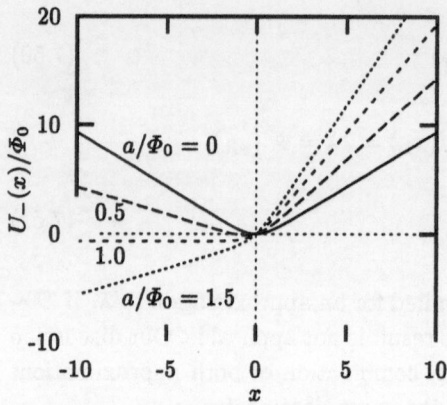

Fig. 7.3. The drift potential (7.55) for various values of a/Φ_0.

the diffusion of a free particle near a soft wall. The corresponding partner potentials read

$$V_{\pm}(x) = \frac{1}{2}\left[\Phi_0^2 + a^2 - \frac{\Phi_0(\Phi_0 \mp D)}{\cosh^2 x} + 2a\Phi_0 \tanh x\right], \tag{7.56}$$

which is the so-called non-symmetric Rosen–Morse potential [RoMo32]. The eigenvalue problem of the associated Schrödinger equation has been studied in various ways [RoMo32, Nie78, JuIn86, BaInWi87].

For obtaining approximate decay rates of not exactly solvable potentials one can choose the same techniques as in the good-SUSY case. For example, for a high barrier of a metastable potential U_- the smallest decay rate $\lambda_0 > 0$ will be very close to zero. Hence, for ϕ_0^- we may take the same ansatz as before,

$$\phi(x) = Ce^{U_-(|x|)/D}\int_{|x|}^{\infty} \mathrm{d}z\, e^{-2U_-(z)/D}, \tag{7.57}$$

and arrive at

$$\lambda_0 \simeq \tfrac{1}{2}C^2D^2. \tag{7.58}$$

Again the normalization constant C may be evaluated via the Laplace method which yields the result

$$\lambda_0 \simeq \frac{D}{2\pi}\sqrt{U_-''(x_{\min})|U_-''(x_{\max})|}$$
$$\times \exp\left\{-\frac{2}{D}\left[U_-(x_{\max}) - U_-(x_{\min})\right]\right\}. \tag{7.59}$$

As before $x_{\min}$ and $x_{\max}$ denote the positions of the local minimum and maximum of U_-, respectively. The above result differs from that of the bistable potential by a factor two which is due to the fact that the metastable potential has only one instead of two local minima.

For lower potential barriers or for the other decay rates one may also apply the WKB approximation (7.50) or the qc-SUSY approximation for broken SUSY

$$\int_{x_L}^{x_R} dx \sqrt{2\lambda - \Phi^2(x)} = D\pi \left(n + \tfrac{1}{2}\right). \tag{7.60}$$

Finally, let us point out that in the Fokker–Planck equation one may allow for a singular drift coefficient, which produces a cusp-like barrier in the drift potential. As an example, let us consider the following drift.

$$\Phi(x) := -\frac{\mathrm{sgn}\, x}{\sqrt{|x|}} - x^2, \qquad U_\pm(x) = \pm \left(\frac{1}{3} x^3 + \sqrt{|x|}\right). \tag{7.61}$$

For the shape of the drift potentials see Fig. 7.4. Obviously, these metastable potentials do not have a stationary distribution and, hence, SUSY is broken. However, they have not necessarily identical decay rates. Mathematically, this could happen because then the superalgebra holds only formally. The corresponding partner Hamiltonians are only formal operators and one has to define carefully their proper domains. See also our remark in point c) of Sect. 5.1. Physically, the system with U_- may decay more rapidly than that with U_+. This is because, once the particle has reached the top of the cusp-like barrier of U_- with a finite velocity it will never return into the well. In the case of U_+, however, because of the flatness of the barrier, the particle has a good change to move back into the well. Hence, we expect U_+ to have a smaller decay rate λ_0^+ then U_-. For an overview on these recent results see [HäTaBo90].

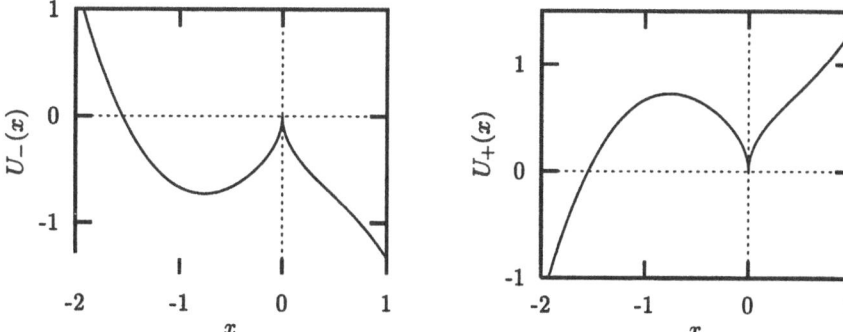

Fig. 7.4. The metastable drift potentials (7.61) showing a cusp-like barrier and well due to a singularity in the SUSY potential.

8. Supersymmetry
in the Pauli and Dirac Equation

The first examples of supersymmetric quantum systems we have mentioned in Sect. 2.1.1 are the Pauli and Dirac Hamiltonian. Here we will discuss some of the implications of supersymmetry for systems which are characterized by these Hamiltonians. In particular, we will consider the Pauli paramagnetism of a two-and three-dimensional non-interacting electron gas. The (IDOS regularized) Witten index is shown to be related to the zero-temperature magnetization of such systems. It is also shown that breaking of SUSY gives rise to counterexamples of the so-called paramagnetic conjecture.

Because of the graded Hilbert space SUSY is a rather natural symmetry of spin-$\frac{1}{2}$ systems. In fact, supersymmetric quantum mechanics has originally been devised for spin systems in statistical mechanics [Nic76]. Nowadays SUSY is discussed, for example, in connection with systems of strongly correlated electrons. Here a supersymmetric extension of the Hubbard model was proposed [EsKoScho90, EsKoScho93] for which an explicit expression for the ground-state wave function can be given. See also [BaJa94, BrGoLiZh95] for other supersymmetric extensions of integrable quantum chains or [GiDoFiReSe90] for an application to fractional statistics. Another field of statistical physics, where supersymmetric methods are used extensively, is that of disordered systems [Efe82, BoEf82, Efe83]. Here SUSY provides an alternative to the replica trick. As recent applications of the SUSY structure in Dirac's Hamiltonian we mention heterojunctions of semiconductors. Supersymmetry clearly shows that there exist localized states at the junction which are not spin degenerate [PaPaVo87, Pan87].

8.1 Pauli's Hamiltonian in Two and Three Dimensions

According to our discussion in Sect. 2.1.1 the Pauli Hamiltonian in three space dimension, which characterizes a spin-$\frac{1}{2}$ particle with charge e and mass m, possesses an $N = 1$ SUSY if the gyromagnetic factor of this particle is equal to two. The self-adjoint supercharge and the Pauli Hamiltonian read

$$Q_1 := \frac{1}{\sqrt{4m}} \left(p - \frac{e}{c} A \right) \cdot \sigma,$$

$$H_{\mathrm{P}}^{(3)} := 2Q_1^2 = \frac{1}{2m} \left(p - \frac{e}{c} A \right)^2 - \frac{e\hbar}{2mc} B \cdot \sigma, \tag{8.1}$$

and act on states in the Hilbert space $\mathcal{H} = L^2(\mathbb{R}^3) \otimes \mathbb{C}^2$. Let us recall that σ is a three-dimensional vector with components given by the 2×2 Pauli matrices σ_i, $i = 1, 2, 3$. Furthermore, the magnetic field is given by the curl of the vector potential $A : \mathbb{R}^3 \to \mathbb{R}^3$, $B := \nabla \times A$. Here we note that because of $N = 1$ we cannot construct a Witten operator analogous to (2.25). Indeed, in the general case none of the Pauli matrices σ_i commutes with $H_{\mathrm{P}}^{(3)}$ because of the Zeemann term. Due to the property $W^2 = 1$ the Witten operator has to be represented by a linear combination of these σ_i's. Consequently, the result derived for $N = 2$ SUSY cannot be applied to the general three-dimensional Pauli Hamiltonian. However, it is still possible to introduce a grading of the Hilbert space. In fact, with the help of the gauge-invariant velocity operator

$$v := \dot{r} = \frac{i}{\hbar} \left[H_{\mathrm{P}}^{(3)}, r \right] = \frac{1}{m} \left(p - \frac{e}{c} A \right) \tag{8.2}$$

and its property

$$(v \cdot \sigma)^2 = 4mQ_1^2 = 2mH_{\mathrm{P}}^{(3)} \tag{8.3}$$

one can introduce the helicity operator

$$\Lambda := \frac{v \cdot \sigma}{\sqrt{2mH_{\mathrm{P}}^{(3)}}} = \operatorname{sgn} Q_1 \tag{8.4}$$

as an alternative to the missing Witten parity. Note that

$$\Lambda^\dagger = \Lambda, \qquad \Lambda^2 = 1, \qquad \left[\Lambda, H_{\mathrm{P}}^{(3)} \right] = 0 \tag{8.5}$$

and, therefore, one can grade the Hilbert space into subspaces of positive and negative helicity. Note, however, that Λ commutes with the supercharge $Q_1 = \sqrt{\frac{1}{2} H_{\mathrm{P}}^{(3)}} \, \Lambda$ and, therefore, leaves the helicity eigenspaces invariant. It does not generate transformations between these spaces. Because of the last relation in (8.5) eigenstates of the Pauli Hamiltonian $H_{\mathrm{P}}^{(3)}$ with strictly positive eigenvalue $E > 0$ may simultaneously be chosen to be an eigenstates of the helicity operator, too:

$$H_{\mathrm{P}}^{(3)} |\psi_E^\pm\rangle = E |\psi_E^\pm\rangle, \qquad \Lambda |\psi_E^\pm\rangle = \pm |\psi_E^\pm\rangle. \tag{8.6}$$

Hence, all strictly positive eigenvalues of $H_{\mathrm{P}}^{(3)}$ are twofold degenerate.

If the magnetic field is chosen such that $B = B(x_1, x_2) e_3$ we may, however, introduce a Witten parity[1] by setting $W := \sigma_3$. This operator now commutes

---

[1] Another class of magnetic fields which gives rise to a Witten parity consists of those with a definite space parity $B(-r) = \pm B(r)$ [GeKr85]. Here, in essence, the Witten operator is given by the space-parity operator Π [cf. eq. (3.30)]. Also for the case of a magnetic monopole field it has been shown that the Pauli-Hamiltonian possesses a dynamical $OSp(1,1)$ supersymmetry [D'HoVi84].

with $H_P^{(3)}$. Note that such a magnetic field, being perpendicular to the (x_1, x_2)-plane, is generated by a vector potential of the form

$$A := \begin{pmatrix} a_1(x_1, x_2) \\ a_2(x_1, x_2) \\ 0 \end{pmatrix}, \qquad B(x_1, x_2) := \frac{\partial a_2}{\partial x_1} - \frac{\partial a_1}{\partial x_2}, \tag{8.7}$$

where $a_i : \mathbb{R}^2 \to \mathbb{R}$, $i = 1, 2$. The three-dimensional Pauli Hamiltonian can be expressed in terms of the two-dimensional one

$$H_P^{(3)} = H_P^{(2)} + \frac{p_3^2}{2m}, \tag{8.8}$$

where

$$H_P^{(2)} := \frac{1}{2m} \sum_{i=1}^{2} \left(p_i - \frac{e}{c} a_i \right)^2 - \frac{e\hbar}{2mc} B\sigma_3. \tag{8.9}$$

It is this two-dimensional Pauli Hamiltonian in a perpendicular magnetic field which possesses an $N = 2$ SUSY. The associated complex supercharge may be defined by

$$Q := \frac{1}{\sqrt{2m}} \left[\left(p_1 - \frac{e}{c} a_1 \right) - i \left(p_2 - \frac{e}{c} a_2 \right) \right] \otimes (\sigma_1 + i\sigma_2) \tag{8.10}$$

and obeys the superalgebra

$$Q^2 = 0, \qquad \{Q, Q^\dagger\} = H_P^{(2)}. \tag{8.11}$$

Because of $W = \sigma_3$ the eigenstates of the Witten operator with positive (negative) Witten parity are spin-up (spin-down) eigenstates. Hence, as a result of SUSY all positive eigenvalues of $H_P^{(2)}$ are spin degenerate. These degenerate eigenstates are related via the SUSY transformation (2.37) or (2.39). Note that the generalized annihilation operator reads in this model

$$A := \frac{1}{\sqrt{2m}} \left[\left(p_1 - \frac{e}{c} a_1 \right) - i \left(p_2 - \frac{e}{c} a_2 \right) \right] \tag{8.12}$$

and the Hamiltonians restricted to the spin-up and spin-down subspace, respectively, are given by

$$H_P^{(2)} \big\lceil \mathcal{H}^+ = AA^\dagger, \qquad H_P^{(2)} \big\lceil \mathcal{H}^- = A^\dagger A. \tag{8.13}$$

Under some mild conditions (B is assumed to be bounded with compact support [CyFrKiSi87]) it has been shown by Aharonov and Casher [AhCa79] that for the two-dimensional Pauli Hamiltonian SUSY is always good if the magnetic flux through the (x_1, x_2)-plane,

$$F := \int_{\mathbb{R}^2} dx_1 dx_2 \, B(x_1, x_2), \tag{8.14}$$

is sufficiently large. To be more precise, the degeneracy of the zero-energy eigenvalue of $H_{\mathrm{P}}^{(2)}$ is given by

$$d := \left[\!\!\left[\frac{|eF|}{2\pi\hbar c} \right]\!\!\right], \tag{8.15}$$

where $[\![z]\!]$ denotes the largest integer which is strictly less than z and $[\![0]\!] = 0$. In other words, the magnitude $|F|$ of the flux (8.14) has to be larger than one flux-quantum $2\pi\hbar c/|e|$ for the existence of a zero-energy eigenstate. The d degenerate eigenstates are all spin-up (spin-down) states for $\mathrm{sgn}(eF) = 1$ $(\mathrm{sgn}(eF) = -1)$. Let us note that the above degeneracy is related to the Witten index by

$$\Delta = -d\,\mathrm{sgn}(eF). \tag{8.16}$$

In concluding this section, we remark that one may also consider the supercharge

$$\widetilde{Q} := \frac{1}{\sqrt{2m}} \left[\left(p_1 - \frac{e}{c}\,a_1 \right) + \mathrm{i} \left(p_2 - \frac{e}{c}\,a_2 \right) \right] \otimes (\sigma_1 + \mathrm{i}\sigma_2) \tag{8.17}$$

which differs from that in (8.10) by the sign in front of the first imaginary unit. As a consequence, one arrives at a Pauli Hamiltonian

$$\widetilde{H}_{\mathrm{P}}^{(2)} := \frac{1}{2m} \sum_{i=1}^{2} \left(p_i - \frac{e}{c}\,a_i \right)^2 + \frac{e\hbar}{2mc}\,B\sigma_3, \tag{8.18}$$

which differs from that in (8.9) by the sign in front of the Zeemann term. Note that $H_{\mathrm{P}}^{(2)}$ and $\widetilde{H}_{\mathrm{P}}^{(2)}$ are not related by a charge conjugation ($e \to -e$). They are related by a reflection at the (x_1, x_2)-plane ($x_3 \to -x_3$). As it stands $\widetilde{H}_{\mathrm{P}}^{(2)}$ characterizes the same particle as $H_{\mathrm{P}}^{(2)}$ but with a gyromagnetic factor $g = -2$, that is, $g \to -g$.

8.2 Pauli Paramagnetism of Non-Interacting Electrons, Revisited

Here we will investigate the implications of the SUSY structure in Pauli's Hamiltonian for the paramagnetic magnetization of a system of non-interacting electrons in two and three dimensions. This magnetization arises from the magnetic moment associated with the spin of the electrons. Moreover, we will confine ourselves to zero temperature, that is, we are looking at the ground state of non-interacting fermions obeying Pauli's exclusion principle. Without interaction the ground-state properties of the electron system are characterized by the single-electron Hamiltonian $H_{\mathrm{P}}^{(2)}$ and $H_{\mathrm{P}}^{(3)}$, respectively. For both values of the space dimension we assume that the magnetic field is perpendicular to the (x_1, x_2)-plane. Since we are not interested in diamagnetic effects due to the orbital motion, the magnetization M of $\mathcal{N} = \mathcal{N}_+ + \mathcal{N}_-$ electrons can be written as

$$M := \mu_B(\mathcal{N}_+ - \mathcal{N}_-), \tag{8.19}$$

where $\mathcal{N}_+$ and $\mathcal{N}_-$ is the number of electrons with spin up and spin down, respectively, in the ground state of the electron system, and

$$\mu_B := \frac{e\hbar}{2mc}. \tag{8.20}$$

is Bohr's magneton.

8.2.1 Two-Dimensional Electron Gas

According to the Pauli principle the ground state of non-interacting electrons with single-electron Hamiltonian $H_P^{(2)}$ is characterized by the reduced single-electron density operator $\Theta(\varepsilon_F - H_P^{(2)})$, where the *Fermi energy* ε_F can be determined from the "normalization" condition

$$\mathrm{Tr}\left(\Theta\left(\varepsilon_F - H_P^{(2)}\right)\right) = \mathcal{N}. \tag{8.21}$$

The magnetization is then given as

$$M = \mu_B \,\mathrm{Tr}\left(\sigma_3 \Theta\left(\varepsilon_F - H_P^{(2)}\right)\right). \tag{8.22}$$

Let us note that because of $W = \sigma_3$ the right-hand-side of (8.22) can be interpreted as a regularized Witten index. In other words, the magnetization is given by the IDOS regularized index (2.48)

$$M = -\mu_B \tilde{\Delta}(\varepsilon_F), \tag{8.23}$$

where the SUSY Hamiltonian in (2.48) is replaced by $H_P^{(2)}$.

For a purely discrete spectrum and a finite degeneracy of each eigenvalue the operator $H_P^{(2)}$ is Fredholm and the regularized index $\tilde{\Delta}(\varepsilon)$ becomes identical to Witten's index:

$$M = -\mu_B \Delta = \mu_B d \,\mathrm{sgn}(eF). \tag{8.24}$$

This is the expected result. Because of SUSY we know that all positive eigenvalues are spin degenerate. Hence, the contribution to the trace in (8.22) of the corresponding eigenstates cancel each other and only the degeneracy of the ground-state energy of $H_P^{(2)}$ contributes. In other words, the Pauli paramagnetism stems from these unpaired zero-energy states only. Because of the SUSY pairing of the excited states only a fraction of all of the electrons contributes to the magnetization. We have illustrated these facts in Fig. 8.1.

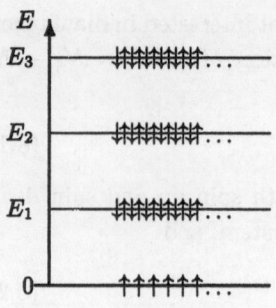

Fig. 8.1. Typical ground state for the two-dimensional electron system. Only the zero-energy states of the single-electron Hamiltonian $H_P^{(2)}$ are unpaired and contribute to the paramagnetic magnetization.

As an example, let us consider the case of a constant magnetic field $B(x_1, x_2) = B > 0$. In order to have well-defined quantities we restrict the configuration space (sample) to a large but finite region with area ℓ^2 such that boundary effects, nevertheless, may be neglected. The flux (8.14) is then given by

$$F = B\ell^2. \tag{8.25}$$

Note that in this case the spectrum of $H_P^{(2)}$ is discrete and consists of (approximately) the well-known equidistant Landau levels $E_n = n\frac{\hbar|e|B}{mc}$. The magnetization per area reads

$$\frac{M}{\ell^2} = \mu_B \frac{eB}{2\pi\hbar c}, \tag{8.26}$$

which is independent of the electron density. Of course, we have assumed that there are sufficiently many electrons in the sample to fill up the zero-energy eigenstates. The paramagnetic susceptibility per unit area then reads

$$\chi := \frac{1}{\ell^2}\frac{\partial M}{\partial B} = \mu_B \frac{e}{2\pi\hbar c} = \frac{e^2}{4\pi mc^2}, \tag{8.27}$$

which is the expected result [Isi91, Isi93].

We note that besides the regularization by a finite region ℓ^2 we have also neglected possible unpaired spins at the Fermi energy. These unpaired spins can only occur when the Fermi energy coincides with an eigenvalue of $H_P^{(2)}$. That is, if we have not completely filled Landau levels.

Despite the fact that we have confined ourselves to the discussion of zero-temperature effects, let us briefly indicate the extension of the above results to finite inverse temperature $\beta < \infty$. Being non-interacting fermions, the magnetization of the electrons at finite temperature reads

$$M(\beta) = \mu_B \text{Tr}\left(\frac{\sigma_3}{1 + \exp\{\beta(H_P^{(2)} - \mu)\}}\right), \tag{8.28}$$

where the chemical potential $\mu = \mu(\beta, \mathcal{N})$ is determined by the "normalization" condition

$$\text{Tr}\left(\frac{1}{1 + \exp\{\beta(H_{\mathrm{P}}^{(2)} - \mu)\}}\right) = \mathcal{N}. \tag{8.29}$$

For vanishing magnetic field μ can be expressed in terms of the Fermi energy $\varepsilon_{\mathrm{F}} = \mu(\infty, \mathcal{N}) \geq 0$ [Lee89, Lee95]

$$\exp\{\beta\mu\} = \exp\{\beta\varepsilon_{\mathrm{F}}\} - 1. \tag{8.30}$$

Assuming, as before, that $H_{\mathrm{P}}^{(2)}$ is Fredholm the contributions of the positive energy eigenstates to the magnetization cancel each other due to SUSY. Hence, we arrive at

$$M(\beta) = \mu_{\mathrm{B}} d\, \mathrm{sgn}(eF)\left(1 + e^{-\beta\mu}\right)^{-1}. \tag{8.31}$$

Assuming further that for an infinitesimal constant magnetic field $B > 0$ the chemical potential is approximately given by the zero-field expression (8.30) the zero-field susceptibility at finite temperature reads

$$\chi_0(\beta) := \lim_{B \to 0} \frac{1}{\ell^2}\frac{\partial M(\beta)}{\partial B} = \frac{e^2}{4\pi mc^2}\left(1 - e^{-\beta\varepsilon_{\mathrm{F}}}\right). \tag{8.32}$$

This result may be interpreted as follows. The fraction of zero-energy states which is not occupied by electrons due to thermal excitation (to available states at the Fermi energy) is given by $e^{-\beta\varepsilon_{\mathrm{F}}}$. Hence, for the magnetization and the susceptibility these electrons do not contribute.

8.2.2 Three-Dimensional Electron Gas

In this section we will utilize our previous results by considering only the case $B = B(x_1, x_2)e_3$ for the three-dimensional problem. Again we will neglect the interaction of the electrons and, hence, have to consider the Hamiltonian $H_{\mathrm{P}}^{(3)}$ as given in (8.8). Obviously, in each subspace of the Hilbert space with a fixed eigenvalue $\hbar k_3$ of p_3 we have the SUSY structure of the two-dimensional Pauli Hamiltonian. Hence, from each of these subspaces we have a contribution to the zero-temperature magnetization as given in the above section. In order to find the total magnetization we simply have to count the occupied eigenvalues of k_3. Assuming a finite range ℓ_3 for the x_3 coordinate the possible eigenvalues are given by

$$k_3 = \frac{2\pi}{\ell_3}n, \qquad n \in \mathbb{Z}. \tag{8.33}$$

However, only those states are occupied with $|k_3|$ less than the *Fermi momentum*

$$k_{\mathrm{F}} := \frac{\sqrt{2m\varepsilon_{\mathrm{F}}}}{\hbar}. \tag{8.34}$$

For large ℓ_3 the number of these occupied states is approximately given by $\ell_3 k_F / \pi$. Multiplying the result (8.23) by this factor we arrive at the three-dimensional magnetization at zero temperature

$$M = -\mu_B \widetilde{\Delta}(\varepsilon_F) \frac{k_F \ell_3}{\pi}. \tag{8.35}$$

Note that here ε_F is the Fermi energy of the three-dimensional system, that is, in (8.21) $H_P^{(2)}$ has to be replaced by $H_P^{(3)}$. The IDOS regularized index, however, is to be calculated with $H_P^{(2)}$.

For the particular case of a constant magnetic field the magnetization and paramagnetic susceptibility per unit volume at zero temperature read

$$\frac{M}{\ell^2 \ell_3} = \mu_B \frac{eB}{2\pi^2 \hbar c} k_F = \mu_B^2 B \frac{m k_F}{\hbar^2 \pi^2},$$
$$\chi = \mu_B^2 \frac{m k_F}{\hbar^2 \pi^2} = \frac{e^2}{4\pi^2 mc^2} k_F. \tag{8.36}$$

Note that in contrast to the two-dimensional case, the three-dimensional zero-temperature paramagnetic susceptibility depends on the electron density and the magnetic field B via the Fermi momentum and has no physical dimension.

In the three-dimensional case practically always exist two eigenvalues of p_3 for which the corresponding eigenstates of $H_P^{(3)}$ coincide with the Fermi energy. Hence, unpaired spins of such states near ε_F are certainly neglected. The above result (8.36) coincides in the limit of vanishing magnetic field with that given in textbooks [AsMe76, Whi83, Isi91].

8.2.3 The Paramagnetic Conjecture and SUSY

Another consequence of SUSY is that it may provide a counterexample to the paramagnetic conjecture due to Hogreve, Schrader and Seiler [HoSchrSe78]. This conjecture states that the ground-state energy of the general Pauli Hamiltonian with an additional scalar potential $V : \mathbb{R}^3 \mapsto \mathbb{R}$,

$$H_P^{(3)}(A, V) := \frac{1}{2m} \left(p - \frac{e}{c} A \right)^2 - \frac{e\hbar}{2mc} B \cdot \sigma + V, \tag{8.37}$$

is always less or equal to that with zero magnetic field:

$$\inf \operatorname{spec} \left(H_P^{(3)}(A, V) \right) \leq \inf \operatorname{spec} \left(H_P^{(3)}(0, V) \right). \tag{8.38}$$

A proof of this inequality exist for arbitrary scalar potential V and a magnetic field being essentially of the form $B = B(x_1^2 + x_2^2) e_3$ [AvSe79]. However, in the general case Avron and Simon [AvSi79] found one counterexample. Nevertheless, it is believed, see [CyFrKiSi87] p.131, that (8.38) "... still holds for general A and selected sets of V ...". Here we note that for the particular case $V = 0$ the factorizability of $H_P^{(3)}(A, 0) = 2Q_1^2 \geq 0$ implies the inequality

$$\text{inf spec} \left(H_{\text{P}}^{(3)}(\boldsymbol{A}, 0) \right) \geq \text{inf spec} \left(H_{\text{P}}^{(3)}(0, 0) \right) = 0. \tag{8.39}$$

This inequality is in the opposite direction of the paramagnetic conjecture (8.38). For arbitrary magnetic fields such that SUSY is a good symmetry the inequality (8.39) can be replaced by an equality. However, for any magnetic field which does break SUSY we have a strict inequality in (8.39) and thereby a counterexample to the conjecture (8.38). Or vice versa, for any magnetic field, for which the paramagnetic conjecture with $V = 0$ can be proven, SUSY will be a good symmetry and hence equality holds in (8.39). The question whether there exist $\boldsymbol{B}$-fields such that for $H_{\text{P}}^{(3)}(\boldsymbol{A}, 0)$ SUSY is broken remains open.

8.3 The Dirac Hamiltonian and SUSY

As already mentioned in Sect. 2.1.1, there is a close connection between SUSY and relativistic quantum systems characterized by the Dirac Hamiltonian. A detailed analysis of SUSY in Dirac's equation is given in the textbook of Thaller [Tha92]. Here we will present only the basic ideas.

Definition 8.3.1 ([Tha92]). *A Dirac Hamiltonian acting on $L^2(\mathbb{R}^3) \otimes \mathbb{C}^4$ and which may be represented in the form*

$$H_{\text{D}} = \begin{pmatrix} M_+ & Q^\dagger \\ Q & -M_- \end{pmatrix} \tag{8.40}$$

with $Q, Q^\dagger, M_+$ and M_- being operators acting on $L^2(\mathbb{R}^3) \otimes \mathbb{C}^2$ is said to be supersymmetric *if the following relations are valid*

$$Q^\dagger M_- = M_+ Q^\dagger, \qquad Q M_+ = M_- Q. \tag{8.41}$$

Note that this definition implies for supersymmetric Dirac Hamiltonians

$$H_{\text{D}}^2 = \begin{pmatrix} Q^\dagger Q + M_+^2 & 0 \\ 0 & Q Q^\dagger + M_-^2 \end{pmatrix}. \tag{8.42}$$

An example, which we have already mentioned in Sect. 2.1.1, is Dirac's Hamiltonian for a magnetic field:

$$H_{\text{D}} := c\boldsymbol{\alpha} \cdot \left(\boldsymbol{p} - \frac{e}{c} \boldsymbol{A} \right) + \beta m c^2. \tag{8.43}$$

The 4×4 Dirac matrices $\boldsymbol{\alpha} = (\alpha_1, \alpha_2, \alpha_3)$ and β close the Dirac algebra

$$\{\alpha_i, \alpha_k\} = 2\delta_{ik}, \qquad \{\alpha_i, \beta\} = 0, \qquad \beta^2 = 1, \tag{8.44}$$

for all $i, k \in \{1, 2, 3\}$. In the standard Pauli-Dirac representation

$$\alpha = \begin{pmatrix} 0 & \sigma \\ \sigma & 0 \end{pmatrix}, \qquad \beta = \begin{pmatrix} 1 & 0 \\ 0 & -1 \end{pmatrix}, \tag{8.45}$$

the Hamiltonian (8.43) is supersymmetric upon the identification

$$Q := c \left(p - \frac{e}{c} A \right) \cdot \sigma = Q^\dagger, \qquad M_\pm := mc^2. \tag{8.46}$$

Note that the above supercharge Q is up to a factor $c/\sqrt{4m}$ identical to Q_1 of Pauli's Hamiltonian (8.1).

An important property of supersymmetric Dirac operators is, that they can be diagonalized,

$$U H_\mathrm{D} U^\dagger = \begin{pmatrix} \sqrt{Q^\dagger Q + M_+^2} & 0 \\ 0 & -\sqrt{QQ^\dagger + M_-^2} \end{pmatrix} \tag{8.47}$$

and, therefore, positive- and negative-energy solutions are decoupled. The unitary transformation U is explicitly given by [Tha88, Tha91, Tha92]

$$U := a_+ + \tau \, \mathrm{sgn}(\mathcal{Q}) \, a_-, \qquad a_\pm := \sqrt{\frac{1}{2} \pm \frac{M}{2|H_\mathrm{D}|}},$$

$$\tau := \begin{pmatrix} 1 & 0 \\ 0 & -1 \end{pmatrix}, \qquad \mathcal{Q} := \begin{pmatrix} 0 & Q^\dagger \\ Q & 0 \end{pmatrix}, \tag{8.48}$$

$$M := \begin{pmatrix} M_+ & 0 \\ 0 & M_- \end{pmatrix}.$$

Here we note that the operators $Q^\dagger Q$ and $QQ^\dagger$ are essential iso-spectral. Consequently, the positive and negative eigenvalues of H_D are closely related. In particular, for $M_+ = M_- = mc^2 > 0$ the spectrum of H_D is symmetric about zero with possible exceptions at $\pm mc^2$ and has a gap from $-mc^2$ to $+mc^2$. The value $+mc^2$ $(-mc^2)$ belongs to the spectrum of H_D if $Q^\dagger Q$ $(QQ^\dagger)$ has zero eigenvalue. That is, if SUSY is good.

8.3.1 Dirac Hamiltonian with a Scalar Potential

As a simple but instructive example we will discuss the problem of the Dirac Hamiltonian with a scalar potential [Tha92] $\beta \Phi_0$, $\Phi_0 : \mathbb{R}^3 \mapsto \mathbb{R}$:

$$H_\mathrm{D} := c\alpha \cdot p + \beta \left(mc^2 + \Phi_0(r) \right). \tag{8.49}$$

In the "supersymmetric" representation [Tha92]

$$\alpha = \begin{pmatrix} 0 & \sigma \\ \sigma & 0 \end{pmatrix}, \qquad \beta = \begin{pmatrix} 0 & -i \\ i & 0 \end{pmatrix} \tag{8.50}$$

this Hamiltonian is supersymmetric with

$$Q := c\boldsymbol{p} \cdot \boldsymbol{\sigma} + \mathrm{i}\left(mc^2 + \varPhi_0(\boldsymbol{r})\right), \qquad M_+ = M_- = 0. \tag{8.51}$$

The spectrum of H_{D} can be obtained from the spectrum of the operators $(\varPhi(\boldsymbol{r}) := mc^2 + \varPhi_0(\boldsymbol{r}))$,

$$\begin{aligned} Q^\dagger Q &= c^2\boldsymbol{p}^2 + \varPhi^2 + \hbar c\boldsymbol{\sigma} \cdot \boldsymbol{\nabla}\varPhi, \\ QQ^\dagger &= c^2\boldsymbol{p}^2 + \varPhi^2 - \hbar c\boldsymbol{\sigma} \cdot \boldsymbol{\nabla}\varPhi, \end{aligned} \tag{8.52}$$

according to (8.47). For further simplification we assume that $\varPhi$ depends only on one coordinate, say x_3, $\varPhi = \varPhi(x_3)$. Such a model describes, for example, a x_3-dependent valence- and conduction-band edge of semiconductors near the $\varGamma$ or L point in the Brillouin zone. The position dependent $\varPhi$ characterizes a position dependent band gap. Examples for such materials are $\mathrm{Pb}_{1-x}\mathrm{Sn}_x\mathrm{Te}$, $\mathrm{Pb}_{1-x}\mathrm{S}_x\mathrm{Se}$, $\mathrm{Hg}_{1-x}\mathrm{Cd}_x\mathrm{Te}$. In particular, all these semiconductors show band inversion, that is, conduction and valence band interchange [DoNiSchl83]. A typical shape for the potential $\varPhi$, which characterizes a $\mathrm{Pb}_{1-x}\mathrm{Sn}_x\mathrm{Te}$ junction, is shown in Fig. 8.2.

With this assumption the above operators simplify to

$$\left.\begin{aligned} Q^\dagger Q \\ QQ^\dagger \end{aligned}\right\} = c^2 p_1^2 + c^2 p_2^2 + H^{(\pm)}, \tag{8.53}$$

where

$$H^{(\pm)} := c^2 p_3^2 + \varPhi^2(x_3) \pm \hbar c\varPhi'(x_3)\sigma_3 \tag{8.54}$$

is a pair of SUSY Hamiltonians acting on $L^2(\mathbb{R}) \otimes \mathbb{C}^2$. In fact, the Hamiltonians are of the type of Witten's model and differ only in the overall sign of the

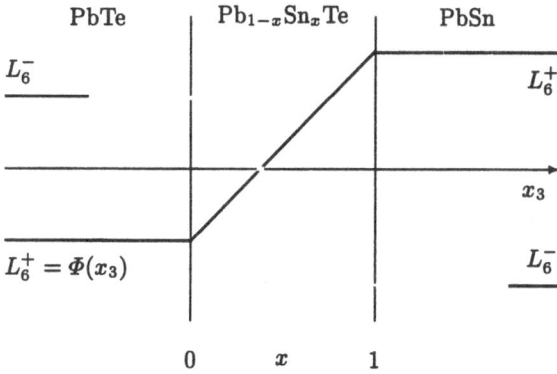

Fig. 8.2. The structure of the $L_6^\pm$ band gap of a $\mathrm{Pb}_{1-x}\mathrm{Sn}_x\mathrm{Te}$ junction as a function of the composition parameter x, which itself varies linearly with x_3 along the junction. For details see [DoNiSchl83]. The scalar potential $\varPhi$ may be identified with the L_6^+ band.

SUSY potential Φ. Our results on the Witten model can immediately be taken over upon replacing $1/\sqrt{2m}$ by c. For the shape of Φ as given in Fig. 8.2, that is, having one zero, $H^{(\pm)}$ both have zero-energy eigenstates. They are given by

$$\phi_0^{(\pm)}(x_3) = C \exp\left\{\mp \hbar c \int_0^{x_3} dz\, \Phi(z)\right\} \chi_{\pm} \tag{8.55}$$

with spinors

$$\chi_+ := \begin{pmatrix} 1 \\ 0 \end{pmatrix}, \qquad \chi_- := \begin{pmatrix} 0 \\ 1 \end{pmatrix}. \tag{8.56}$$

In other words, these semiconductors have unpaired spin-up and spin-down states in the conduction and valence band, respectively, which are localized at the junction [PaPaVo87, Pan87]. Finally, let us note that for $\Phi(x) = mc^2 \tanh x$ the spectrum and eigenfunctions of H_D can explicitly be calculated due to shape invariance as we have shown in Sect. 5.2. Of course, any other shape-invariant SUSY potential will also allow for an exact solution of the eigenvalue problem of H_D [CoKhMuWi88]. For example, a detailed discussion of the radial Dirac equation for the Coulomb problem is given in Chap. 7.4 of Thaller's book [Tha92]. See also [Suk85d, JaSt86].

9. Concluding Remarks and Overview

In the last fifteen years supersymmetric quantum mechanics has increasingly been utilized as an elegant and effective method in various branches of theoretical physics. In the present book we have discussed some of the recent developments in supersymmetric quantum mechanics. Only a small number of applications in quantum and statistical physics have been treated. In Table 9.1 we give a list[1] of applications in quantum mechanics, mathematical, statistical, condensed-matter, atomic, and nuclear physics. It appears that the Witten model, which we have extensively studied, is the prototype of a supersymmetric model in quantum mechanics. Despite the fact that this model is rather simple, it finds applications, for example, in the study of classical stochastic dynamical systems. Even for the understanding of semiconductor heterojunctions the Witten model can be utilized. A detailed analysis of its classical and quasi-classical properties has revealed many interesting results. The most impressive one is the quasi-classical supersymmetric quantization condition derived and discussed in Chap. 6. This approximation leads to exact bound-state spectra of all (via the factorization method) completely solvable problems. In this respect, it is superior to the well-established WKB approximation.

Besides the Witten model we have also discussed the supersymmetric nature of Pauli's Hamiltonian and its implications to magnetic properties of electronic systems. In fact, supersymmetry becomes more and more a basic tool in the theory of such systems with or without interactions. In connection with a quasi-classical approximation analogous to that of the Witten model it might be possible to derive additional information about their spectral properties in one, two or more dimensions. In this context also disordered systems are of interest, in particular, because of their relevance for the understanding of the quantized Hall effect.

Further developments which can be based on the present work may be performed for classical systems with stochastic dynamics. For instance, a detailed

---

[1] This list is not meant to be complete. Also the given references are certainly fragmentary.

functional analytic investigation of Witten's model for singular SUSY potentials may provide additional insights into the diffusion problem over cusp-like barriers. Let us also mention that in classical systems driven by colored noise supersymmetry may be utilized for a further study of their dynamics.

The Table 9.1 does not contain particle and high-energy physics. There exist many textbooks covering these topics. Below we given an incomplete list:

1) S.J. Gates, M.T. Grisaru, M. Roček and W. Siegel, *Superspace or One Thousand and One Lessons in Supersymmetry*, (Benjamin/Cummings, London, 1983)

2) P.G.O. Freund, *Introduction to Supersymmetry*, (Cambridge University Press, London, 1986)

3) R.N. Mohapatra, *Unification and Supersymmetry*, (Springer-Verlag, New York, 1986)

4) O. Piguet and K. Sibold, *Renormalized Supersymmetry*, (Birkhäuser, Boston, 1986)

5) P. West, *Introduction to Supersymmetry and Supergravity*, (World Scientific, Singapore, 1986)

6) M.B. Green, J.H. Schwarz, and E. Witten, *Superstring Theory*, Vols. 1 and 2, (Cambridge University Press, Cambridge, 1987)

7) H.J.W. Müller-Kirsten and A. Wiedemann, *Supersymmetry: An Introduction with Conceptual and Calculational Details*, (World Scientific, Singapore, 1987)

8) F. Gieres, *Geometry of Supersymmetric Gauge Theories*, Lecture Notes in Physics 302, (Springer-Verlag, Berlin, 1988)

9) M. Kaku, *Introduction to Superstrings*, (Springer-Verlag, New York, 1988)

10) P. van Nieuwenhuizen, *Anomalies in Quantum Field Theories: Cancellation of Anomalies in $d = 10$ Supergravity*, (Leuven University Press, Leuven, 1988)

11) M. Müller, *Consistent Classical Supergravity Theories*, Lecture Notes in Physics 336, (Springer-Verlag, Berlin, 1989)

12) J. Lopusanski, *Introduction to Symmetry and Supersymmetry in Quantum Field Theory*, (World Scientific, Singapore, 1991)

13) L. Castellani, R. D'Auria and P. Fré, *Supergravity and Superstrings*, (World Scientific, Singapore, 1991)

14) J. Wess and J. Bagger, *Supersymmetry and Supergravity*, (Princeton University Press, Princeton, 1991), 2nd ed.

Table 9.1. Some applications of SUSY in theoretical physics

Quantum mechanics	References
solvable potentials, factorization method	Sect. 5.2[a]
quasi-exactly solvable potentials	[JaKuKh89]
quasi-classical approximation	[CoBaCa85, InJu93a], Chap. 6[a]
tunneling	[KeKoSu88], [RoRoRo91, CoKhSu95][a]
variational approach	[GoReTh93]
δ-expansion method	[CoRo90], [CoKhSu95][a]
large-N expansion	[ImSu85], [RoRoRo91, CoKhSu95][a]
bounds on ground-state energy	[Schm85]
level ordering	[BaGrMa84], [Gro91][a]
Pauli Hamiltonian	[GeKr85][a], Sect. 8.1[a]
Dirac Hamiltonian	[Tha92][a], Sect. 8.3[a]
inverse scattering method	[Suk85c]
coherent states	[FuAi93]
supercoherent states	[BaSchmBa88, BaSchmHa89, FaKoNiTr91]
supercoherent-state path integrals	[Koc96]
Berry and other phases	[IiKu87, BhDuBaKh94, IlKaMe95]
integrability, many-body problems	[ShSu93]
SUSY breaking, instantons	[Wit81, SaHo82, KuLiMü-Ki93]
Korteweg-de Vries equation, solitons	[Tha92][a], [KuLiMü-Ki93]
fractional SUSY	[Dur93a, Dur93b, AzMa96]
anyons	[Sen92, RoTa92]
quantum chaos	[Guh95]
parasupersymmetry	[BeDe90], [CoKhSu95][a]
orthosupersymmetry	[CoKhSu95][a]
q-deformation	[IlUz93, Deb93, BoDa93]
SUSY and geometric motion	[BoWeRi93]
SUSY on Riemann surfaces	[BoIl94, DoIl94]
Mathematical physics	
pseudoclassical mechanics	[Cas76c, LaDoGu93], Chap. 4[a]
Morse theory	[Wit82b, CyFrKiSi87]
Atiyah-Singer index theorem	[Alv83, FrWi84]
localization techniques	[BlTh95, SchwZa95][a]

[a] and references therein.

Table 9.1. *cont.*

Statistical and condensed matter physics	References
Fokker–Planck equation	Sect. 7.1[a]
Langevin equation	Sect. 7.2[a]
Nicolai map	[Nic80b, EzKl85]
Colored noise	[LeMaRi87]
Grassmann Brownian motion	[Rog87, Cor93]
random walks	[Jau90, RoRe95]
Pauli paramagnetism	[Jun95], Sect. 8.2
supersymmetric Hubbard models	[EsKoScho90, EsKoScho93]
integrable quantum chains	[BaJa94, BrGoLiZh95]
fractional statistics	[GiDoFiReSe90]
random matrices	[Zuk94]
disordered systems	[Efe83]
classical diffusion	[BoCoGeDo90][a]
electron localization-delocalization	[ToZaPi88, CoDeMo95][a]
density of states	[Weg83, BrGrIt84], [FiHa90][a]
conductivity	[HaJo91][a]
correlation function	[Tka94]
Jahn-Teller systems	[JaSt84]
lattice models	[Sha85, BeIg95]
semiconductor heterojunctions	[PaPaVo87], Sect. 8.3.1

Atomic physics	References
alkali-metal atoms	[KoNi84, KoNi85a]
quantum-defect theory	[KoNi85b, KoNiTr88]
Stark effect	[BlKo93]
Penning trap	[Kos93]
Rydberg atoms, superrevivals	[BlKo94, BlKo95], [BlKoPo96][a]

Nuclear physics	References
dynamical supersymmetry	[Iac80], [Ver87][a]
classification of spectra	[BaBaIa81]
nuclear-nuclear potential	[UrHe83]
collective excitations	[BaReGe86]
superdeformed nuclei	[AmBiCaDe91]

[a] and references therein.

References

[AbMo80]   P.B. Abraham and H.E. Moses,
                Changes in potentials due to changes in the point spectrum:
                Anharmonic oscillators with exact solutions,
                Phys. Rev. A **22** (1980) 1333-1340

[AdDuKhSu88]   R. Adhikari, R. Dutt, A. Khare, and U.P. Sukhatme,
                Higher-order WKB approximation in supersymmetric
                quantum mechanics,
                Phys. Rev. A **38** (1988) 1679-1686

[AhCa79]   Y. Aharonov and A. Casher,
                Ground state of a spin-1/2 charged particle in a two-
                dimensional magnetic field,
                Phys. Rev. A **19** (1979) 2461-2462

[AkCo84]   R. Akhoury and A. Comtet,
                Anomalous behavior of the Witten index – Exactly soluble
                models,
                Nucl. Phys. **B246** (1984) 253-278

[AlBr95]   S. Albeverio and Z. Brzeźniak,
                Oscillatory integrals on Hilbert spaces and Schrödinger
                equation with magnetic fields,
                J. Math. Phys. **36** (1995) 2135-2156

[AlHo76]   S. Albeverio and R.J. Høegh-Krohn,
                Mathematical Theory of Feynman Path Integrals,
                Lecture Notes in Mathematics **523**,
                (Springer-Verlag, Berlin, 1976)

[Alv83]   L. Alvarez-Gaumé,
                Supersymmetry and the Atiyah-Singer index theorem,
                Commun. Math. Phys. **90** (1983) 161-173

[AmBiCaDe91]   R.D. Amado, R. Bijker, F. Cannata, and J.P. Dedonder,
                Supersymmetric quantum mechanics and superdeformed
                nuclei,
                Phys. Rev. Lett. **67** (1991) 2777-2779

[AnBoIo84]   A.A. Andrianov, N.V. Borisov, and M.V. Ioffe,
The factorization method and quantum systems with equivalent energy spectra,
Phys. Lett. **105A** (1984) 19-22

[Arn73]   L. Arnold
Stochastische Differentialgleichungen,
(Oldenbourg Verlag, München, 1973)

[AsMe76]   N.W. Ashcroft and N.D. Mermin,
Solid State Physics,
(Holt, Rinehart and Winston, New York, 1976)

[AtBoPa73]   M. Atiyah, R. Bott, and V.K. Patodi,
On the heat equation and the index theorem,
Inventiones math. **19** (1973) 279-330

[AtPaSi75]   M.F. Atiyah, V.K. Padoti, and I.M. Singer,
Spectral asymmetry and Riemannian geometry. I,
Math. Proc. Cambrigde Phil. Soc. **77** (1975) 43-69

[AvSe79]   J.E. Avron and R. Seiler,
Paramagnetism for nonrelativistic electrons and Euclidean Dirac particles,
Phys. Rev. Lett. **42** (1979) 931-943

[AvSi79]   J. Avron and B. Simon,
A counterexample to the paramagnetic conjecture,
Phys. Lett. **75A** (1979) 41-42

[AzMa96]   J.A. de Azcárraga and A.J. Macfarlane,
Group theoretical foundations of fractional supersymmetry,
J. Math. Phys. **37** (1996) 1115-1127

[BaDeJa91]   M. Baake, R. Delbourgo, and P.D. Jarvis,
Models for supersymmetric quantum mechanics,
Aust. J. Phys. **44** (1991) 353-362

[BaReGe86]   M. Baake, P. Reinicke, and A. Gelberg,
A Hamiltonian with broken supersymmetry for the xenon isotopes,
Phys. Lett. **166B** (1986) 10-17

[BaJa94]   T.H. Baker and P.D. Jarvis,
Quantum superspin chains,
Int. J. Mod. Phys. B **8** (1994) 3623-3635

[BaBaIa81]   A.B. Balantekin, I. Bars, and F. Iachello,
U(6/4) supersymmetry in nuclei,
Nucl. Phys. **A370** (1981) 284-316

[BaSchmBa88]   A.B. Balantekin, H.A. Schmitt, and B.R. Barrett,
                Coherent states for the harmonic oscillator representations
                of the orthosymplectic supergroup Osp(1/2N,R),
                J. Math. Phys. **29** (1988) 1634-1639

[BaSchmHa89]   A.B. Balantekin, H.A. Schmitt, and P. Halse,
                Coherent states for the noncompact supergroups
                Osp(2/2N,R),
                J. Math. Phys. **30** (1989) 274-279

[BaDuGaKhPaSu93] D.T. Barclay, R. Dutt, A. Gangopadhyaya, A. Khare,
                A. Pagnamenta, and U.P. Sukhatme,
                New exactly solvable Hamiltonians: Shape invariance and
                self-similarity,
                Phys. Rev. A **48** (1993) 2786-2797

[BaKhSu93]   D.T. Barclay, A. Khare, and U. Sukhatme,
                Is the lowest order supersymmetric WKB approximation
                exact for all shape invariant potentials?,
                Phys. Lett. A **183** (1993) 263-266

[BaMa91]   D.T. Barclay and C.J. Maxwell,
                Shape invariance and the SWKB series,
                Phys. Lett. A **157** (1991) 357-360

[BaBoCa81]   A. Barducci, F. Bordi, and R. Casalbuoni,
                Path integral quantization of spinning particles interacting
                with crossed external electromagnetic fields,
                Nuovo Cim. **64B** (1981) 287-315

[BaCaLu76]   A. Barducci, R. Casalbuoni, and L. Lusanna,
                Supersymmetries and pseudoclassical relativistic electron,
                Nuovo Cim. **35A** (1976) 377-399

[Bar86]   A.O. Barut,
                Electron theory and path integral formulation of quantum-
                electrodynamics from a classical action,
                in M.C. Gutzwiller, A. Inomata, J.R. Klauder, and L. Streit eds.,
                Path Integrals from meV to MeV,
                Bielefeld Encounters in Physics and Mathematics VII,
                (World Scientific, Singapore, 1986) p.381-95

[BaInWi87]   A.O. Barut, A. Inomata, and R. Wilson,
                Algebraic treatment of second Pöschl-Teller, Morse-Rosen
                and Eckart equations,
                J. Phys. A **20** (1987) 4083-4096

[BaRo92]   A.O. Barut and P. Roy,
                The embedding of supersymmetry into the dynamical groups,
                in A. Frank, T. Seligman, and K. Wolf eds.,
                Group Theory in Physics,
                AIP Conference Proceedings 266,
                (American Institute of Physics, New York, 1992) p.248-254

[BaZa84]   A.O. Barut and N. Zanghi,
                Classical model of the Dirac electron,
                Phys. Rev. Lett. **52** (1984) 2009-2012

[Bau91]   B. Baumgartner,
                *Perturbations of supersymmetric systems in quantum
                mechanics,*
                in A. Boutet de Monvel, P. Dita, G. Nenciu, and R. Purice eds.,
                Recent Developments in Quantum Mechanics,
                Mathematical Physics Studies Nr. 12
                (Kluwer Acad. Publ., Dordrecht, 1991) p.195-208

[BaGrMa84]   B. Baumgartner, H. Grosse, and A. Martin
                The Laplacian of the potential and the order of energy levels,
                Phys. Lett. **146B** (1984) 363-366

[BaSp94]   D. Baye and J.-M. Sparenberg,
                *Most general form of phase-equivalent radial potentials for
                arbitrary modifications of the bound spectrum,*
                Phys. Rev. Lett. **73** (1994) 2789-2792

[BeDe90]   J. Beckers and N. Debergh,
                Parastatistics and supersymmetry in quantum mechanics,
                Nucl. Phys. **B340** (1990) 767-776

[BeIg95]   B. Berche and F. Iglói,
                *Realization of supersymmetric quantum mechanics in
                inhomogeneous Ising models,*
                J. Phys. A **28** (1995) 3579-3590

[Ber66]   F.A. Berezin,
                The Method of Second Quantization,
                (Academic Press, New York, 1966)

[BeMa75]   F.A. Berezin and M.S. Marinov,
                Classical spin and Grassmann algebra,
                JETP Lett. **21** (1975) 320-321

[BeMa77]   F.A. Berezin and M.S. Marinov,
                *Particle spin dynamics as the Grassmann variant of classical
                mechanics,*
                Ann. Phys. (NY) **104** (1977) 336-362

[BeBr84]   M. Bernstein and L.S. Brown,
                Supersymmetry and the bistable Fokker–Planck equation,
                Phys. Rev. Lett. **52** (1984) 1933-1935

[BhDuBaKh94]   D. Bhaumik, B. Dutta-Roy, B.K. Bagchi, and A. Khare,
                Berry phase for supersymmetric shape-invariant potentials,
                Phys. Lett. A **193** (1994) 11-14

[BlTh95]   M. Blau and G. Thompson,
                  *Localization and diagonalization: A review of functional
                  integral techniques for low-dimensional gauge theories and
                  topological field theories*,
                  J. Math. Phys. **36** (1995) 2192-2236

[BlKo93]   R. Bluhm and V.A. Kostelecký,
                  Atomic supersymmetry and the Stark effect,
                  Phys. Rev. A **47** (1993) 794-808

[BlKo94]   R. Bluhm and V.A. Kostelecký,
                  *Atomic supersymmetry, Rydberg wave packets, and radial
                  squeezed states*,
                  Phys. Rev. A **49** (1994) 4628-4640

[BlKo95]   R. Bluhm and V.A. Kostelecký,
                  *Long-term evolution and revival structure of Rydberg wave
                  packets for hydrogen and alkali-metal atoms*,
                  Phys. Rev. A **51** (1995) 4767-4786

[BlKoPo96]   R. Bluhm, V.A. Kostelecký, and J.A. Porter,
                  *The evolution and revival structure of localized quantum wave
                  packets*,
                  Am. J. Phys. **64** (1996) 944-954

[BoEf82]   T. Bohr and K.B. Efetov,
                  *Derivation of Green function for disordered chain by
                  integrating over commuting and anticommuting variables*,
                  J. Phys. C **15** (1982) L249-L254

[BoGeGrSchwSi87] D. Bollé, F. Gesztesy, H. Grosse, W. Schweiger, and
                  B. Simon,
                  *Witten index, axial anomaly, and Krein's spectral shift
                  function in supersymmetric quantum mechanics*,
                  J. Math. Phys. **28** (1987) 1512-1525

[BoDa93]   D. Bonatsos and C. Daskaloyannis,
                  *General deformation schemes and $N = 2$ supersymmetric
                  quantum mechanics*,
                  Phys. Lett. B **307** (1993) 100-105

[BoIl94]   N.V. Borisov and K.N. Ilinski,
                  *$N = 2$ supersymmetric quantum mechanics on Riemann
                  surfaces with meromorphic superpotentials*,
                  Commun. Math. Phys. **161** (1994) 177-194

[BoCoGeDo90]   J.P. Bouchaud, A. Comtet, A. Georges, and P. Le Doussal,
                  *Classical diffusion of a particle in a one-dimensional random
                  force field*,
                  Ann. Phys. (NY) **201** (1990) 285-341

130   References

[BoWeRi93]   L.J. Boya, R.F. Wehrhahn, and A. Rivero,
Supersymmetry and geometric motion,
J. Phys. A **26** (1993) 5824-5834

[BoBl84]   D. Boyanovsky and R. Blankenbecler,
Fractional indices in supersymmetric theories,
Phys. Rev. D **30** (1984) 1821-1824

[BrGoLiZh95]   A.J. Bracken, M.D. Gould, J.R. Links, and Y.-Z. Zhang,
*A new supersymmetric and exactly solvable model of
correlated electrons*,
Phys. Rev. Lett. **74** (1995) 2768-2771

[BrGrIt84]   E. Brézin, D.J. Gross, and C. Itzykson,
*Density of states in the presence of a strong magnetic field
and a random impurity*,
Nucl. Phys. **B235** (1984) 24-44

[Bri26]   L. Brillouin,
*La méchanique ondulatoire de Schrödinger; une méthode
générale de résolution par approximations successives*,
Comptes Rendus Acad. Sci. (Paris) **183** (1926) 24-26

[BrDeZuVeHo76]   L. Brink, S. Deser, B. Zumino, P. Di Veccia and, P. Howe,
Local supersymmetry for spinning particles,
Phys. Lett. (1976) **64B** 435-438

[Cal78]   C. Callias,
Axial anomalies and index theorems on open spaces,
Comm. Math. Phys. **62** (1978) 213-234

[CaDeWi95]   P. Cartier and C. DeWitt-Morette
A new perspective on functional integration,
J. Math. Phys. **36** (1995) 2237-2312

[Cas91]   J. Casahorran,
*A family of supersymmetric quantum mechanics models with
singular superpotentials*,
Phys. Lett. A **156** (1991) 425-428

[Cas76a]   R. Casalbuoni,
Relativity and supersymmetries,
Phys. Lett. **62B** (1976) 49-50

[Cas76b]   R. Casalbuoni,
On the quantization of systems with anticommuting variables,
Nuovo Cim. **33A** (1976) 115-125

[Cas76c]   R. Casalbuoni,
The classical mechanics for Bose–Fermi systems,
Nuovo Cim. **33A** (1976) 389-431

[CeGi83]   S. Cecotti and L. Girardello,
                Stochastic and parastochastic aspects of supersymmetric
                functional measures: A new non-perturbative approach to
                supersymmetry,
                Ann. Phys. (NY) **145** (1983) 81-99

[CoMa67]   S. Coleman and J. Mandula,
                All possible symmetries of the S matrix,
                Phys. Rev. **159** (1967) 1251-1256

[CoBaCa85]   A. Comtet, A.D. Bandrauk, and D.K. Campbell,
                Exactness of semiclassical bound state energies for
                supersymmetric quantum mechanics,
                Phys. Lett. B **150** (1985) 159-162

[CoDeMo95]   A. Comtet, J. Desbois, and C. Monthus,
                Localization properties in one dimensional disordered
                supersymmetric quantum mechanics,
                Ann. Phys. (NY) **239** (1995) 312-350

[CoGr94]   F. Constantinescu and H.F. de Groote,
                Geometrische und algebraische Methoden der Physik:
                Supermannigfaltigkeiten und Virasoro-Algebren,
                (B.G. Teubner, Stuttgart, 1994)

[CoFr83]   F. Cooper and B. Freedman,
                Aspects of supersymmetric quantum mechanics,
                Ann. Phys. (NY) **146** (1983) 262-288

[CoGiKh87]   F. Cooper, J.N. Ginocchio, and A. Khare,
                Relationship between supersymmetry and solvable potentials,
                Phys. Rev. D **36** (1987) 2458-2473

[CoKhMuWi88]   F. Cooper, A. Khare, R. Musto, and A. Wipf,
                Supersymmetry and the Dirac equation,
                Ann. Phys. (NY) **187** (1988) 1-28

[CoKhSu95]   F. Cooper, A. Khare, and U. Sukhatme,
                Supersymmetry and quantum mechanics,
                Phys. Rep. **251** (1995) 267-385

[CoRo90]   F. Cooper and P. Roy,
                δ *expansion for the superpotential,*
                Phys. Lett. A **143** (1990) 202-206

[Cor93]   R. Corns,
                Path integrals in $N = 2$ supersymmetric quantum mechanics,
                J. Math. Phys. **34** (1993) 2723-2741

[Cor89]   J.F. Cornwell,
                Group Theory in Physics, Vol. III: Supersymmetries and
                Infinite-Dimensional Algebras,
                (Academic Press, London, 1989)

[CrRi83]   M. de Crombrugghe and V. Rittenberg,
                 Supersymmetric quantum mechanics,
                 Ann. Phys. (NY) **151** (1983) 99-126

[CyFrKiSi87]   H.L. Cycon, R.G. Froese, W. Kirsch, and B. Simon,
                 *Schrödinger Operators with Application to Quantum
                 Mechanics and Global Geometry*,
                 (Springer-Verlag, Berlin, 1987)

[Dar1882]   G. Darboux,
                 Sur une proposition relative aux équations linéaires,
                 Comptes Rendus Acad. Sci. (Paris) **94** (1882) 1456-1459

[Deb93]   N. Debergh,
                 *On a q-deformation of the supersymmetric Witten
                 model*,
                 J. Phys. A **26** (1993) 7219-7226

[Dei78]   P.A. Deift,
                 Applications of a commutation formula,
                 Duke Math. J. **45** (1978) 267-310

[deLaRa91]   O.L. deLange and R.E. Raab,
                 Operator Methods in Quantum Mechanics,
                 (Claredon Press, Oxford, 1991)

[DeWit92]   B. DeWitt,
                 Supermanifolds, Second Edition
                 (Cambridge University Press, Cambridge, 1992)

[DeWiMaNe79]   C. DeWitt-Morette, A. Maheshwari, and B. Nelson,
                 Path integration in non-relativistic quantum mechanics,
                 Phys. Rep. **50** (1979) 255-372

[D'HoVi84]   E. D'Hoker and L. Vinet,
                 *Supersymmetry of the Pauli equation in the presence of a
                 magnetic monopole*,
                 Phys. Lett. **137B** (1984) 72-76

[Dir64]   P.A.M. Dirac,
                 Lectures on Quantum Mechanics,
                 (Belfer Graduate School of Science, Yeshiva University,
                 New York, 1964)

[DoIl94]   A.D. Dolgallo and K.N. Ilinski,
                 *Generalized supersymmetric quantum mechanics on Riemann
                 surfaces with meromorphic superpotentials*,
                 J. Math. Phys. **35** (1994) 2074-2082

[DoNiSchl83]   R. Dornhaus, G. Nimtz, and B. Schlicht,
                 Narrow-Gap Semiconductors,
                 Springer Tracts in Modern Physics Vol. 98,
                 (Springer-Verlag, Berlin, 1983)

[Dun31]   J.L. Dunham,
               *The Wentzel-Brillouin-Kramers method of solving the wave
               equation,*
               Phys. Rev. **41** (1931) 713-720

[Dur93a]   S. Durand,
               Fractional supersymmetry and quantum mechanics,
               Phys. Lett. B **312** (1993) 115-120

[Dur93b]   S. Durand,
               Fractional superspace formulation of generalized mechanics,
               Mod. Phys. Lett. A **8** (1993) 2323-2334

[DuKhSu86]   R. Dutt, A. Khare, and U.P. Sukhatme,
               *Exactness of supersymmetric WKB spectra for shape-
               invariant potentials,*
               Phys. Lett. B **181** (1986) 295-298

[Eck86]   B. Eckhardt,
               Maslov-WKB theory for supersymmetric Hamiltonians,
               Phys. Lett. **168B** (1986) 245-247

[Efe82]   K.B. Efetov,
               Supersymmetry method in localized theory,
               Sov. Phys. JETP **55** (1982) 514-521

[Efe83]   K.B. Efetov,
               Supersymmetry and theory of disordered metals,
               Adv Phys. **32** (1983) 53-127

[Ein05]   A. Einstein,
               *Über die von der molekularkinetischen Theorie der Wärme
               geforderte Bewegung von in ruhenden Flüssigkeiten
               suspendierten Teilchen,*
               Ann. Physik (Leipzig) **17** (1905) 549-560

[Ein06]   A. Einstein,
               Zur Theorie der Brownschen Bewegung,
               Ann. Physik (Leipzig) **19** (1906) 371-381

[Eng88]   M.J. Englefield,
               Exact solutions of a Fokker–Planck equation,
               J. Stat. Phys. **52** (1988) 369-381

[EsKoScho90]   F.H.L. Essler, V.E. Korepin, and K. Schoutens,
               *New exactly solvable model of strongly correlated electrons
               motivated by high-T_c superconductivity,*
               Phys. Rev. Lett. **68** (1990) 2960-2963

[EsKoScho93]   F.H.L. Essler, V.E. Korepin, and K. Schoutens,
               Electronic model for superconductivity,
               Phys. Rev. Lett. **70** (1993) 73-76

[Exn85]   P. Exner,
                   Open Quantum Systems and Feynman Integrals,
                   (D. Reidel Publ., Dordrecht, 1985)

[EzKl85]   H. Ezawa and J.R. Klauder,
                   Fermions without fermions, — The Nicolai map revisited —,
                   Prog. Theor. Phys. **74** (1985) 904-915

[Fad63]   L.D. Faddeyev,
                   The inverse problem in the quantum theory of scattering,
                   J. Math. Phys. **4** (1963) 72-104

[FaKoNiTr91]   B.W. Fatyga, V.A. Kotelecký, M.M. Nieto, and D.R. Truax,
                   Supercoherent states,
                   Phys. Rev. D **43** (1991) 1403-1412

[FeTs82]   M.V. Feĭgel'man and A.M. Tsvelik,
                   Hidden supersymmetry of stochastic dissipative dynamics,
                   Sov. Phys. JETP **56** (1982) 823-830

[FeFuDe79]   G. Feldman, T. Fulton, and A. Devoto
                   Energy levels and level ordering in the WKB approximation,
                   Nucl. Phys. **B154** (1979) 441-462

[Fel87]   B. Felsager,
                   Geometry, Particles and Fields,
                   (Odense University Press, 1987) 4th ed.

[Fey48]   R.P. Feynman,
                   Space-time approach to non-relativistic quantum mechanics,
                   Rev. Mod. Phys. **20** (1948) 367-387

[Fey61]   R.P. Feynman,
                   Quantum Electrodynamics,
                   (W.A. Benjamin Inc., New York, 1961), p.50

[FeHi65]   R.P. Feynman and A.R. Hibbs,
                   Quantum Mechanics and Path Integrals,
                   (McGraw-Hill, New York, 1965)

[FiHa90]   H.J. Fischbeck and R. Hayn,
                   *On the density of states of a disordered Peierls–Fröhlich
                   chain*,
                   phys. stat. sol. (b) **158** (1990) 565-572

[FiLeMü95]   W. Fischer, H. Leschke, and P. Müller,
                   *The functional-analytic versus the functional-integral
                   approach to quantum Hamiltonians:
                   The one-dimensional hydrogen atom*,
                   J. Math. Phys. **36** (1995) 2313-2323

[Fok14]   A.D. Fokker,
                 Die mittlere Energie rotierender elektrischer Dipole im
                 Strahlungsfeld,
                 Ann. Physik (Leipzig) **43** (1914) 810-820

[FrBaHaUz88]   S.H. Fricke, A.B. Balantekin, P.J. Hatchell, and T. Uzer,
                 Uniform semiclassical approximation to supersymmetric
                 quantum mechanics,
                 Phys. Rev. A **37** (1988) 2797-2804

[FrWi84]   D. Friedan and P. Windey,
                 Supersymmetric derivation of the Atiyah–Singer index and
                 the chiral anomaly,
                 Nucl. Phys. **B235** (1984) 395-416

[Fuc86]   J. Fuchs,
                 Physical state conditions and supersymmetry breaking in
                 quantum mechanics,
                 J. Math. Phys. **27** (1986) 349-353

[FuAi93]   T. Fukui and N. Aizawa,
                 Shape-invariant potentials and an associated coherent state,
                 Phys. Lett. A **180** (1993) 308-313

[GaPaSu93]   A. Gangopadhyaya, P.K. Panigrahi, and U.P. Sukhatme,
                 Supersymmetry and tunneling in an asymmetric double well,
                 Phys. Rev. A **47** (1993) 2720-2724

[Gar90]   C.W. Gardiner,
                 Handbook of Stochastic Methods for Physics, Chemistry and
                 the Natural Sciences,
                 Springer Series in Synergetics, Vol. 13,
                 (Springer-Verlag, Berlin, 1990) 2nd ed., corrected printing

[GeLe51]   I.M. Gel'fand and B.M. Levitan,
                 On the determination of a differential equation from its
                 spectral function,
                 Am. Math. Soc. Transl. **1** (1951) 253-304

[Gen83]   L.É. Gendenshteîn,
                 Derivation of exact spectra of the Schrödinger equation by
                 means of supersymmetry,
                 JETP Lett. **38** (1983) 356-358

[GeKr85]   L.É. Gendenshteîn and I.V. Krive,
                 Supersymmetry in quantum mechanics,
                 Sov. Phys. Usp. **28** (1985) 645-666

[GiPa77a]   E. Gildner and A. Patrascioiu,
                 Pseudoparticle contributions to the energy spectrum of a one-
                 dimensional system,
                 Phys. Rev. D **16** (1977) 423-430

[GiPa77b]   E. Gildner and A. Patrascioiu,
                 Effect of fermions upon tunneling in a one-dimensional system,
                 Phys. Rev. D **16** (1977) 1802-1804

[GiKoReMa89]   S. Giler, P. Kosiński, J. Rembieliński, and P. Maślanka,
                 Ground-state energy calculation in broken SUSYQM by a WKB-like method,
                 J. Phys. A **22** (1989) 647-661

[GiDoFiReSe90]   S.M. Girvin, A.H. MacDonald, M.P.A. Fisher, S.-J. Rey, and J.P. Sethna,
                 Exactly soluble models of fractional statistics,
                 Phys. Rev. Lett. **65** (1990) 1671-1674

[GoLi71]   Y.A. Gol'fand and E.P. Likhtam,
                 Extension of the algebra of Poincaré group generators and violation of P-invariance,
                 JETP Lett. **13** (1971) 323-326

[Goz83]   E. Gozzi,
                 Ground-state wave-function "representation",
                 Phys. Lett. **129B** (1983) 432-436

[Goz84]   E. Gozzi,
                 Onsager principle of microscopic reversibility and supersymmetry,
                 Phys. Rev. D **30** (1984) 1218-1227

[GoReTh93]   E. Gozzi, M. Reuter, and W.D. Thacker,
                 Variational methods via supersymmetric techniques,
                 Phys. Lett. A **183** (1993) 29-32

[GrRo85]   R. Graham and D. Roekaerts,
                 Supersymmetric quantum mechanics and stochastic processes in curved configuration space,
                 Phys. Lett. **109A** (1985) 436-440

[Gre66]   H.S. Green,
                 Quantenmechanik in algebraischer Darstellung,
                 (Springer-Verlag, Berlin, 1966)

[Gro91]   H. Grosse,
                 Supersymmetric quantum mechanics,
                 in A. Boutet de Monvel, P. Dita, G. Nenciu, and R. Purice eds.,
                 Recent Developments in Quantum Mechanics,
                 Mathematical Physics Studies Nr. 12
                 (Kluwer Acad. Publ., Dordrecht, 1991) p.299-327

[GrMa84]   H. Grosse and A. Martin
                 Two theorems on the level order in potential models,
                 Phys. Lett. **134B** (1984) 368-372

[GrPi87]   H. Grosse and L. Pittner,
                Supersymmetric quantum mechanics defined as sesquilinear
                forms,
                J. Phys. A **20** (1987) 4265-4284

[Guh95]   T. Guhr,
                Transition toward quantum chaos: with supersymmetry from
                Poisson to Gauss,
                preprint, cond-mat/9510052 (1995)

[Gut67]   M.C. Gutzwiller,
                Phase-integral approximation in momentum space and the
                bound states of an atom,
                J. Math. Phys. **8** (1967) 1979-2000

[Gut92]   M.C. Gutzwiller,
                The semi-classical quantization of chaotic Hamiltonian
                systems,
                in M.-J. Giannoni, A. Voros, and J. Zinn-Justin eds.,
                chaos and quantum physics, Les Houches 1989,
                (North Holland, Amsterdam, 1991) p.201-249

[HaLoSo75]   R. Haag, J.T. Łopuszański, and M. Sohnius,
                All possible generators of supersymmetries of the S-matrix,
                Nucl. Phys. **B88** (1975) 257-274

[HäTaBo90]   P. Hänggi, P. Talkner, and M. Borkovec,
                Reaction-rate theory: fifty years after Kramers,
                Rev. Mod. Phys. **62** (1990) 251-341

[HaJo91]   R. Hayn and W. John,
                Instanton approach to the conductivity of a disordered solid,
                Nucl. Phys. **B348** (1991) 766-786

[HeTe92]   M. Henneaux and C. Teitelboim,
                Quantization of Gauge Systems,
                (Princeton University Press, Princeton, 1992)

[Hir83]   M. Hirayama,
                Supersymmetric quantum mechanics and index theorem,
                Prog. Theor. Phys. **70** (1983) 1444-1453

[HoSchrSe78]   H. Hogreve, R. Schrader, and R. Seiler,
                A conjecture on the spinor functional determinant,
                Nucl. Phys. **B142** (1978) 525-534

[HoZh82]   M.O. Hongler and W.M. Zheng,
                Exact solution for the diffusion in bistable potentials,
                J. Stat. Phys. **29** (1982) 317-327

[HoZh83]   M.-O. Hongler and W.M. Zheng,
                Exact results for the diffusion in a class of asymmetric
                bistable potentials,
                J. Math. Phys. **24** (1983) 336-340

[HuIn48]   T.E. Hull and L. Infeld,
                *The factorization method, hydrogen intensities, and related
                problems*,
                Phys. Rev. **74** (1948) 905-909

[Iac80]   F. Iachello,
                Dynamical supersymmetry in nuclei,
                Phys. Rev. Lett. **44** (1980) 772-775

[IiKu87]   S. Iida and H. Kuratsuji,
                *Phase holonomy, zero-point energy cancellation and
                supersymmetric quantum mechanics*,
                Phys. Lett. B **198** (1987) 221-225

[IlKaMe95]   K.N. Ilinski, G.V. Kalinin, and V.V. Melezhik,
                Berry phase and supersymmetric topological index,
                J. Math. Phys. **36** (1995) 6611-6624

[IlUz93]   K.N. Ilinski and V.M. Uzdin,
                *Quantum superspace, q-extended supersymmetry and
                parasupersymmetric quantum mechanics*,
                Mod. Phys. Lett. A **8** (1993) 2657-2670

[ImSu85]   T. Imbo and U. Sukhatme,
                Supersymmetric quantum mechanics and large-N expansion,
                Phys. Rev. Lett. **54** (1985) 2184-2187

[Inc56]   E.L. Ince,
                Ordinary Differential Equations,
                (Dover Publ., New York, 1956)

[Inf41]   L. Infeld,
                On a new treatment of some eigenvalue problems,
                Phys. Rev. **59** (1941) 737-747

[InHu51]   L. Infeld and T.E. Hull,
                The factorization method,
                Rev. Mod. Phys. **23** (1951) 21-68

[InJu93a]   A. Inomata and G. Junker,
                *Quasi-classical approach to path integrals in supersymmetric
                quantum mechanics*,
                in H. Cerdeira, S. Lundqvist, D. Mugnai, A. Ranfagni,
                V. Sa-yakanit, and L.S. Schulman eds.,
                Lectures on Path Integration: Trieste 1991,
                (World Scientific, Singapore, 1993) p.460-482

[InJu93b]   A. Inomata and G. Junker,
             Quasi-classical approach in supersymmetric quantum mechanics,
             in J.Q. Liang, M.L. Wang, S.N. Qiao, and D.C. Su eds.,
             Proceedings of International Symposium on Advanced Topics of Quantum Physics,
             (Science Press, Beijing, 1993) p.61-74

[InJu94]   A. Inomata and G. Junker,
             Quasiclassical path-integral approach to supersymmetric quantum mechanics,
             Phys. Rev. A **50** (1994) 3638-3649

[InJuSu93]   A. Inomata, G. Junker, and A. Suparmi,
             Remarks on semiclassical quantization rule for broken SUSY,
             J. Phys. A **26** (1993) 2261-2264

[InKuGe92]   A. Inomata, H. Kuratsuji, and C.C. Gerry,
             Path Integrals and Coherent States of SU(2) and SU(1,1),
             (World Scientific, Singapore, 1992)

[InSe95]   K. Intriligator and N. Seiberg,
             Lectures on supersymmetric gauge theories and electric-magnetic duality,
             preprint, hep-th/9509066 (1995)

[Isi91]   A. Isihara,
             Condensed Matter Physics,
             (Oxford University Press, New York, 1991)

[Isi93]   A. Isihara,
             Electron Liquids,
             Springer Series in Solid-State Sciences 96,
             (Springer-Verlag, Berlin, 1993)

[JaLeLe87]   A. Jaffe, A. Lesniewski, and M. Lewenstein,
             Ground state structure in supersymmetric quantum mechanics,
             Ann. Phys. (NY) **178** (1987) 313-329

[Jar89]   P.D. Jarvis,
             Supersymmetric quantum mechanics and the index theorem,
             in M.N. Barber and M.K. Murray eds.,
             Proceedings of the Center for Mathematical Analysis,
             (Australian National University, Volume 22, 1989) p.50-81

[JaSt84]   P.D. Jarvis and G.E. Stedman,
             Supersymmetry in Jahn-Teller systems,
             J. Phys. A **17** (1984) 757-776

[JaSt86]   P.D. Jarvis and G.E. Stedman,
                Supersymmetry in second-order relativistic equations for the
                hydrogen atom,
                J. Phys. A **19** (1986) 1373-1385

[JaKuKh89]   D.P. Jatkar, C.N. Kumar, and A. Khare,
                A quasi-exactly solvable problem without $Sl(2)$ symmetry,
                Phys. Lett. A **142** (1989) 200-202

[Jau88]   H.R. Jauslin,
                Exact propagator and eigenfunction for multistable models
                with arbitrary prescribed N lowest eigenvalues,
                J. Phys. A **21** (1988) 2337-2350

[Jau90]   H.R. Jauslin,
                Supersymmetric partners for random walks,
                Phys. Rev. A **41** (1990) 3407-3410

[JeRo84]   A. Jevicki and J.P. Rodrigues,
                Singular potentials and supersymmetry breaking,
                Phys. Lett. **146B** (1984) 55-58

[Jos67]   A. Joseph,
                Self-adjoint ladder operators (I),
                Rev. Mod. Phys. **39** (1967) 829-837

[Jun95]   G. Junker,
                Recent developments in supersymmetric quantum mechanics,
                Turk. J. Phys. **19** (1995) 230-248 and cond-mat/9403088

[JuIn86]   G. Junker and A. Inomata,
                Path integral on S^3 and its application to the Rosen-Morse
                oscillator,
                in M.C. Gutzwiller, A. Inomata, J.R. Klauder, and L. Streit eds.,
                Path Integrals from meV to MeV,
                Bielefeld Encounters in Physics and Mathematics VII,
                (World Scientific, Singapore, 1986) p.315-334

[JuMa94]   G. Junker and S. Matthiesen,
                Supersymmetric classical mechanics,
                J. Phys. A **27** (1994) L751-L755

[JuMa95]   G. Junker and S. Matthiesen,
                ADDENDUM: Pseudoclassical mechanics and its solution,
                J. Phys. A **28** (1995) 1467-1468

[JuMaIn95]   G. Junker, S. Matthiesen, and A. Inomata,
                Classical and quasi-classical aspects of supersymmetric
                quantum mechanics,
                preprint, hep-th/9510230, to appear in the proceedings of
                the *VII International Conference on Symmetry Methods in*
                Physics, Dubna, July 10–16, 1995

[Kam81]   N.G. van Kampen,
               Stochastic Processes in Physics and Chemistry,
               (North-Holland, Amsterdam, 1981)

[Kat84]   T. Kato,
               Perturbation Theory for Linear Operators,
               (Springer-Verlag, Berlin, 1984) 2nd ed., 2nd corrected printing

[KaMi89]   R.K. Kaul and L. Mizrachi,
               *On non-perturbative contributions to vacuum energy in
               supersymmetric quantum mechanical models*,
               J. Phys. A **22** (1989) 675-685

[KeKoSu88]   W.-Y. Keung, E. Kovacs, and U. Sukhatme,
               Supersymmetry and double-well potentials,
               Phys. Rev. Lett. **60** (1988) 41-44

[Kha85]   A. Khare,
               *How good is the supersymmetry-inspired WKB quantization
               condition?*,
               Phys. Lett. **161B** (1985) 131-135

[KhSu93]   A. Khare and U.P. Sukhatme,
               *New shape-invariant potentials in supersymmetric quantum
               mechanics*,
               J. Phys. A **26** (1993) L901-L904

[KiSaSk85]   A. Kihlberg, P. Salomonson, and B.S. Skagerstam,
               Witten's index and supersymmetric quantum mechanics,
               Z. Phys. C **28** (1985) 203-209

[Kl1872]   F. Klein,
               *Vergleichende Betrachtungen über neuere geometrische
               Forschungen*,
               Programm zum Eintritt in die philosophische Facultät und
               den Senat der k. Friedrich-Alexanders-Universität zu
               Erlangen,
               (Verlag von Andreas Deichert, Erlangen, 1872)

[Kle95]   H. Kleinert,
               *Path Integrals in Quantum Mechanics Statistics and Polymer
               Physics*,
               (World Scientific, Singapore, 1995) 2nd ed.

[Koc96]   E.A. Kochetov,
               Path-integral formalism for the supercoherent states,
               Phys. Lett. A **217** (1996) 65-72

[KoSuIn94]   C.K. Koleci, A. Suparmi, and A. Inomata,
               Semiclassical quantization of the Gendenshtein systems,
               in V. Sa-yakanit, J.-O. Berananda, and W. Sritrakool eds.,
               Path Integrals in Physics,
               (World Scientific, Singapore, 1994) p.127-136

[Kos93]   V.A. Kostelecký,
                  Atomic supersymmetry, oscillators and the Penning trap,
                  in B. Gruber and T. Otsuka eds.,
                  *Symmetries in Science VII: Dynamical Symmetries and
                  Spectrum-Generating Algebras in Physics,*
                  (Plenum, New York, 1993)

[KoCa85]   V.A. Kostelecký and D.K. Campbell eds.,
                  Supersymmetry in Physics,
                  (North-Holland, Amsterdam, 1985)

[KoNi84]   V.A. Kostelecký and M.M. Nieto,
                  *Evidence for a phenomenological supersymmetry in atomic
                  physics,*
                  Phys. Rev. Lett. **53** (1984) 2285-2288

[KoNi85a]   V.A. Kostelecký and M.M. Nieto,
                  *Evidence from alkali-metal-atom transition probability for a
                  phenomenological atomic supersymmetry,*
                  Phys. Rev. A **32** (1985) 1293-1298

[KoNi85b]   V.A. Kostelecký and M.M. Nieto,
                  Analytic wave functions for atomic quantum-defect theory,
                  Phys. Rev. A **32** (1985) 1293-1298

[KoNiTr88]   V.A. Kostelecký, M.M. Nieto, and D.R. Truax,
                  Fine structure and analytic quantum-defect wave functions,
                  Phys. Rev. A **38** (1988) 4413-4418

[Kra26]   M.A. Kramers,
                  Wellenmechanik und halbzahlige Quantisierung,
                  Z. Phys. **39** (1926) 828-840

[Kra40]   H.A. Kramers,
                  *Brownian motion in a field of force and the diffusion model
                  of chemical reactions,*
                  Physica **7** (1940) 284-304

[KuLiMü-Ki93]   D.S. Kulshreshtha, J.-Q. Liang, and H.J.W. Müller-Kirsten,
                  *Fluctuation equations about classical field configurations and
                  supersymmetric quantum mechanics,*
                  Ann. Phys. (NY) **225** (1993) 191-211

[LaRoBa90]   A. Lahiri, P.K. Roy, and B. Bagchi,
                  Supersymmetry in quantum mechanics,
                  Int. J. Mod. Phys. A **5** (1990) 1383-1456

[Lan08]   P. Langevin,
                  Sur la théorie du mouvement brownien,
                  Comptes Rendus Acad. Sci. (Paris) **146** (1908) 530-533

[LaRoTi82]   F. Langouche, D. Roekaerts, and E. Tirapegui
                  Functional Integration and Semiclassical Expansion,
                  (D. Reidel Publ., Dordrecht, 1982)

[LaDoGu93]   A. Lasenby, C. Doran, and S. Gull,
                  *Grassmann calculus, pseudoclassical mechanics, and
                  geometric algebra,*
                  J. Math. Phys. **34** (1993) 3683-3712

[LaMi89]   H.B. Lawson, Jr. and M.-L. Michelsohn,
                  Spin Geometry,
                  (Princeton University Press, Princeton, 1989)

[Lee89]   M.H. Lee,
                  *Chemical potential of a D-dimensional free Fermi gas at finite
                  temperature,*
                  J. Math. Phys. **30** (1989) 1837-1839

[Lee95]   M.H. Lee,
                  *Polylogarithmic analysis of chemical potential and
                  fluctuations in a D-dimensional free Fermi gas at low
                  temperatures,*
                  J. Math. Phys. **36** (1995) 1217-1231

[LeMaRi87]   Th. Leiber, F. Marchesoni, and H. Risken,
                  Colored noise and bistable Fokker–Planck equations,
                  Phys. Rev. Lett. **59** (1987) 1381-1384
                  ERRATUM: Phys. Rev. Lett. **60** (1988) 659

[Lesch81]   H. Leschke,
                  Path integral approach to fluctuations in dynamic processes,
                  in H. Haken ed.,
                  Chaos and Order in Nature,
                  Springer Series in Synergetics, Vol. 11,
                  (Springer-Verlag, Berlin, 1981) p.157-163

[LeSchm77]   H. Leschke and M. Schmutz,
                  *Operator orderings and functional formulations of quantum
                  and stochastic dynamics,*
                  Z. Physik B **27** (1977) 85-94

[Lev89]   G. Lévai,
                  A search for shape-invariant solvable potentials,
                  J. Phys. A **22** (1989) 689-702

[LuPu86]   M. Luban and D.L. Pursey,
                  *New Schrödinger equations for old: Inequivalence of Darboux
                  and Abraham-Moses constructions,*
                  Phys. Rev. D **33** (1986) 431-436

[MaObSo66]   W. Magnus, F. Oberhettinger, and R.P. Soni,
                  *Formulas and Theorem for the Special Functions of
                  Mathematical Physics,*
                  (Springer-Verlag, Berlin, 1966) 3rd ed.

[MaZu86]   J. Mañes and B. Zumino,
                *WKB method, SUSY quantum mechanics and the index
                theorem,*
                Nucl. Phys. **B270** (1986) 651-686

[MaSoZa88]   F. Marchesoni, P. Sodano, and M. Zannetti,
                Supersymmetry and bistable soft potentials,
                Phys. Rev. Lett. **61** (1988) 1143-1146

[Mat95]   S. Matthiesen,
                *Supersymmetrische klassische Mechanik und ihre
                Quantisierung,*
                Diploma thesis,
                (University Erlangen-Nürnberg, 1995)

[Mil68]   W. Miller Jr.,
                Lie Theory and Special Functions,
                (Academic Press, New York, 1968) p.267-276

[Mis91]   S.P. Misra,
                Introduction to Supersymmetry and Supergravity,
                (John Wiley, New York, 1991)

[Miy68]   H. Miyazawa,
                Spinor currents and symmetries of baryons and mesons,
                Phys. Rev. **170** (1968) 1596-1590

[Mor52]   C. Morette,
                *On the definition and approximation of Feynman's path
                integral,*
                Phys. Rev. **81** (1952) 848-852

[MuGoKh95]   N.R. Murali, T.R. Govindarajan, and A. Khare,
                *Exactness of the broken supersymmetric, semiclassical
                quantization rule,*
                Phys. Rev. A **52** (1995) 4259-4261

[Mur89]   Y. Murayama,
                Supersymmetric Wentzel–Kramers–Brillouin (SWKB) method,
                Phys. Lett. A **136** (1989) 455-457

[Nak90]   M. Nakahara,
                Geometry, Topology and Physics,
                (IOP Publishing Ltd., Bristol, 1990)

[Nel64]   E. Nelson,
                Feynman integrals and the Schrödinger equation,
                J. Math. Phys. **5** (1964) 332-343

[Nic76]   H. Nicolai,
                Supersymmetry and spin systems,
                J. Phys. A **9** (1976) 1497-1506

[Nic80a]   H. Nicolai,
                On a new characterization of scalar supersymmetric theories,
                Phys. Lett. **89B** (1980) 341-346

[Nic80b]   H. Nicolai,
                Supersymmetry and functional integration measures,
                Nucl. Phys. **B176** (1980) 419-428

[Nic91]   H. Nicolai,
                Supersymmetrische Quantenmechanik,
                Phys. Blätter **47** (1991) 387-392

[NiWi84]   A.J. Niemi and L.C.R. Wijewardhana,
                Fractionization of the Witten index,
                Phys. Lett. **138B** (1984) 389-392

[Nie78]   M.M. Nieto,
                Exact wave-function normalization for the $B_0 \tanh z -$
                $U_0 \cosh^{-2} z$ and Pöschl–Teller potentials,
                Phys. Rev. A **17** (1978) 1273-1283

[Noe18]   E. Noether,
                Invariante Variationsprobleme,
                Nachr. d. König. Gesellsch. d. Wiss. zu Göttingen,
                Math-phys. Klasse (1918) 235-257

[OlPe83]   M.A. Olshanetsky and A.M. Perelomov,
                Quantum integrable systems related to Lie symmetries,
                Phys. Rep. **94** (1983) 313-404

[PaSu90]   A. Pagnamenta and U. Sukhatme,
                Non-divergent semiclassical wave functions in
                supersymmetric quantum mechanics,
                Phys. Lett. A **151** 7-11

[PaSu93]   P.K. Panigrahi and U.P. Sukhatme,
                Singular superpotentials in supersymmetric quantum
                mechanics,
                Phys. Lett. A **178** (1993) 251-257

[Pan87]   O.A. Pankratov,
                Supersymmetric inhomogeneous semiconductor structures
                and the nature of a parity anomaly in (2+1) electrodynamics,
                Phys. Lett. A **121** (1987) 360-366

[PaPaVo87]   O.A. Pankratov, S.V. Pakhomov, and B.A. Volkov,
                Supersymmetry in heterojunctions: Band-inverting contact
                on the basis of $Pb_{1-x}Sn_x Te$ *and* $Hg_{1-x}Cd_x Te$,
                Solid State Comm. **61** (1987) 93-96

[PaSo79]   G. Parisi and N. Sourlas,
                Random magnetic fields, supersymmetry, and negative
                dimensions,
                Phys. Rev. Lett. **43** (1979) 744-745

[PaSo82]   G. Parisi and N. Sourlas,
              *Supersymmetric field theories and stochastic differential
              equations*,
              Nucl. Phys. **B206** (1982) 321-332

[Pau51]   W. Pauli,
              Ausgewählte Kapitel aus der Feldquantisierung,
              Lecture Notes at the ETH Zürich 1950-1951,
              (Haag, Zürich, 1951)

[Pla17]   M. Planck,
              *Über einen Satz der statistischen Dynamik und seine.
              Erweiterung in der Quantentheorie*,
              Sitzber. Preuss. Akad. Wiss. (1917) 324-341

[Pur86]   D.L. Pursey,
              New families of isospectral Hamiltonians,
              Phys. Rev. D **33** (1986) 1048-1055

[RaSeVa87]   K. Raghunathan, M. Seetharaman, and S.S. Vasan,
              On the exactness of the SUSY semiclassical quantization rule,
              Phys. Lett. B **188** (1987) 351-352

[ReSi75]   M. Reed and B. Simon,
              II Fourier Analysis, Self-Adjointness,
              (Academic Press, San Diego, 1975)

[ReSi80]   M. Reed and B. Simon,
              I Functional Analysis, revised and enlarged edition,
              (Academic Press, San Diego, 1980)

[Ric78]   R.D. Richtmyer,
              Principles of Advanced Mathematical Physics, Vol. I,
              (Springer-Verlag, New York, 1978)

[Ris89]   H. Risken,
              *The Fokker–Planck Equation, Methods of Solution and
              Applications*,
              (Springer-Verlag, Berlin, 1989) 2nd ed.

[Roe94]   G. Roepstorff,
              Path Integral Approach to Quantum Physics,
              (Springer-Verlag, Berlin, 1994)

[Rog87]   A. Rogers,
              *Fermionic path integration and Grassmann Brownian
              motion*,
              Commun. Math. Phys. **113** (1987) 353-368

[RoMo32]   N. Rosen and P.M. Morse,
              On the vibrations of polyatomic molecules,
              Phys. Rev. **42** (1932) 210-217

[RoKr68]   C. Rosenzweig and J.B. Krieger,
              Exact quantization conditions,
              J. Math. Phys. **9** (1968) 849-860

[RoRe95]   H.C. Rosu and M. Reyes,
              Supersymmetric time-continuous discrete random walks,
              Phys. Rev. E **51** (1995) 5112-5115

[RoRoRo91]   B. Roy, P. Roy, and R. Roychoudhury,
              On solutions of quantum eigenvalue problems:
              A supersymmetric approach,
              Fortschr. Phys. **89** (1991) 211-258

[RoTa92]   P. Roy and R. Tarrach,
              Supersymmetric anyon quantum mechanics,
              Phys. Lett. B **274** (1992) 59-64

[SaSt74]   A. Salam and J. Strathdee,
              Super-gauge transformations,
              Nucl. Phys. **B76** (1974) 477-482

[SaHo82]   P. Salomonson and J.W. van Holten,
              Fermionic coordinates and supersymmetry in quantum
              mechanics,
              Nucl. Phys. **B196** (1982) 509-531

[Sche79]   M. Scheunert,
              The Theory of Lie Superalgebras,
              Lecture Notes in Mathematics **716**,
              (Springer-Verlag, Berlin, 1979)

[Schm78]   U. Schmincke,
              On Schrödinger factorization method for Sturm-Liouville
              operators,
              Proc. Roy. Soc. Edinburgh **80A** (1978) 67-84

[Schm85]   M. Schmutz,
              The factorization method and ground state bounds,
              Phys. Lett. **108A** (1985) 195-196

[Schr40]   E. Schrödinger,
              A method of determining quantum-mechanical eigenvalues
              and eigenfunctions,
              Proc. Roy. Irish Acad. **46A** (1940) 9-16

[Schr41a]   E. Schrödinger,
              Further studies on solving eigenvalue problems by
              factorization,
              Proc. Roy. Irish Acad. **46A** (1941) 183-206

[Schr41b]   E. Schrödinger,
              The factorization of the hypergeometric equation,
              Proc. Roy. Irish Acad. **47A** (1941) 53-54

[Schu81]   L.S. Schulman,
                Techniques and Applications of Path Integration,
                (John Wiley, New York, 1981)

[Schw95]   F. Schwabl,
                Quantum Mechanics,
                (Springer-Verlag, Berlin, 1995), 2nd rev. ed.

[SchwZa95]   A. Schwarz and O. Zaboronsky,
                Supersymmetry and localization,
                preprint, hep-th/9511112 (1995)

[Sei95]   N. Seiberg,
                The power of duality – Exact results in 4D SUSY field theory,
                preprint, hep-th/9506077 (1995)

[SeWi94a]   N. Seiberg and E. Witten,
                *Electric-magnetic duality, monopole condensation, and
                confinement in N=2 supersymmetric Yang-Mills theory*,
                Nucl. Phys. **B426** (1994) 19-52,
                ERRATUM: Nucl. Phys. **B430** (1994) 485-486

[SeWi94b]   N. Seiberg and E. Witten,
                *Monopoles, duality and chiral symmetry breaking in N=2
                supersymmetric QCD*,
                Nucl. Phys. **B431** (1994) 484-550

[Sen92]   D. Sen,
                *Some supersymmetric features in the spectrum of anyons in
                a harmonic potential*,
                Phys. Rev. D **46** (1992) 1846-1857

[Sha85]   Y. Shapir,
                Supersymmetric statistical models on the lattice,
                Physica **15D** (1985) 129-137

[ShSu93]   B.S. Shastry and B. Sutherland,
                Super Lax pairs and infinite symmetries in the $1/r^2$ system,
                Phys. Rev. Lett. **70** (1993) 4029-4033

[ShSmVa88]   M.A. Shifman, A.V. Smilga, and A.I. Vainshtein,
                On the Hilbert space of supersymmetric quantum systems,
                Nucl. Phys. **B299** (1988) 79-90

[SiSt86]   L.P. Singh and F. Steiner,
                *Fermionic path integrals, the Nicolai map and the Witten
                index*,
                Phys. Lett. **166B** (1986) 155-159

[SiTePoLu90]   A.N. Sissakian, V.M. Ter-Antonyan, G.S. Pogosyan, and
                I.V. Lutsenko,
                Supersymmetry of a one-dimensional hydrogen atom,
                Phys. Lett. A **143** (1990) 247-249

[Smo06]   M. von Smoluchowski,
            *Zur kinetischen Theorie der Brownschen Molekularbewegung
            und der Suspensionen,*
            Ann. Physik (Leipzig) **21** (1906) 756-780

[Sou85]   N. Sourlas,
            Introduction to supersymmetry in condensed matter physics,
            Physica **15D** (1985) 115-122

[Suk85a]   C.V. Sukumar,
            *Supersymmetry, factorisation of the Schrödinger equation
            and a Hamiltonian hierarchy,*
            J. Phys. A **18** (1985) L57-L61

[Suk85b]   C.V. Sukumar,
            *Supersymmetric quantum mechanics of one-dimensional
            systems,*
            J. Phys. A **18** (1985) 2917-2936

[Suk85c]   C.V. Sukumar,
            *Supersymmetric quantum mechanics and the inverse
            scattering method,*
            J. Phys. A **18** (1985) 2937-2955

[Suk85d]   C.V. Sukumar,
            *Supersymmetry and the Dirac equation for a central Coulomb
            field,*
            J. Phys. A **18** (1985) L697-L701

[Sup92]   A. Suparmi,
            *Semi-classical quantization rules in supersymmetric quantum
            mechanics,*
            Doctoral thesis,
            (State University of New York, Albany, 1992)

[Tha88]   B. Thaller,
            *Normal forms of an abstract Dirac operator and applications
            to scattering theory,*
            J. Math. Phys. **29** (1988) 249-257

[Tha91]   B. Thaller,
            Dirac particles in magnetic fields,
            in A. Boutet de Monvel, P. Dita, G. Nenciu, and R. Purice eds.,
            Recent Developments in Quantum Mechanics,
            Mathematical Physics Studies Nr. 12
            (Kluwer Acad. Publ., Dordrecht, 1991) p.351-366

[Tha92]   B. Thaller,
            The Dirac Equation,
            (Springer-Verlag, Berlin, 1992)

[Tka94]   V.M. Tkachuk,
               The supersymmetry representation for correlation functions
               of disordered systems,
               preprint, ESI-92 (1994),
               available at http://www.esi.ac.at/ESI-Preprints.html

[ToZaPi88]   E. Tosatti, M. Zannetti, and L. Pietronero,
               Exponentiated random walks, supersymmetry and
               localization,
               Z. Phys. B **73** (1988) 161-166

[Tri90]   S. Trimper,
               Supersymmetry breaking for dynamical systems,
               J. Phys. A **23** (1990) L169-L174

[UrHe83]   L.F. Urrutia and E. Hernádez,
               Long-range behavior of nuclear forces as a manifestation of
               supersymmetry in nature,
               Phys. Rev. Lett. **51** (1983) 755-758

[Var92]   Y.P. Varshni,
               Relative convergences of WKB and SWKB approximation,
               J. Phys. A **25** (1992) 5761-5777

[Ver87]   J. Vervier,
               Boson-fermion symmetries and supersymmetries in nuclear
               physics,
               Riv. Nuovo Cim. **10** (1987) N.9 1-102

[Vle28]   J.H. van Vleck,
               The correspondence principle in the statistical interpretation
               of quantum mechanics,
               Proc. Natl. Acad. Sci. U.S.A. **14** (1928) 178-188

[VoVoTo85]   J. Vogel, E. Vogel, and C. Toepffer,
               Relaxation in a Fermi liquid,
               Ann. Phys. (NY) **164** (1985) 463-494

[VoAk72]   D.V. Volkov and V.P. Akulov,
               Is the neutrino a Goldstone particle?,
               Phys. Lett. **46B** (1973) 109-110

[Wae32]   B.L. van der Waerden,
               Die Gruppentheorische Methode in der Quantenmechanik,
               (Springer-Verlag, Berlin, 1932)

[Weg83]   F. Wegner,
               Exact density of states for lowest Landau level in white noise
               potential superfield representation for interacting systems,
               Z. Phys. B **51** (1983) 279-285

[Wei79]   E.J. Weinberg,
               Parameter counting for multimonopole solutions,
               Phys. Rev. D **20** (1979) 936-944

[Wen26]   G. Wenzel,
                 Eine Verallgemeinerung der Quantenbedingungen für die
                 Zwecke der Wellenmechanik,
                 Z. Phys. **38** (1926) 518-529

[WeZu74a]   J. Wess and B. Zumino,
                 Supergauge transformations in four dimensions,
                 Nucl. Phys. **B70** (1974) 39-50

[WeZu74b]   J. Wess and B. Zumino,
                 Supergauge invariant extension of quantum electrodynamics,
                 Nucl. Phys. **B78** (1974) 1-13

[Wey31]   H. Weyl,
                 Gruppentheorie und Quantenmechanik,
                 (Hirzel, Leipzig, 1931)

[Whi83]   R.M. White
                 Quantum Theory of Magnetism,
                 Springer-Series in Solid-State Science, Vol. 32,
                 (Springer-Verlag, Berlin, 1983) 2nd corr. and updated ed.

[Wig31]   E.P. Wigner,
                 Gruppentheorie und ihre Anwendung auf die
                 Quantenmechanik der Atomspektren,
                 (Vieweg, Braunschweig, 1931)

[Wig39]   E. Wigner,
                 On unitary representations of the inhomogeneous Lorentz
                 group,
                 Ann. Math. **40** (1931) 149-204

[Wit81]   E. Witten,
                 Dynamical breaking of supersymmetry,
                 Nucl. Phys. **B188** (1981) 513-554

[Wit82a]   E. Witten,
                 Constraints on supersymmetry breaking,
                 Nucl. Phys. **B202** (1982) 253-316

[Wit82b]   E. Witten,
                 Supersymmetry and Morse theory,
                 J. Diff. Geom. **17** (1982) 661-692

[Wit92]   S. Wittmer,
                 Supersymmetrische Quantenmechanik und semiklassische
                 Näherung,
                 Diploma thesis,
                 (University Stuttgart, 1992)

[WiWe92]   S. Wittmer and U. Weiss,
                 A shape-invariant generalization of the harmonic oscillator
                 in supersymmetric quantum mechanics,
                 preprint, University Stuttgart (1992), unpublished

[ZiJu93]   J. Zinn-Justin,
            Quantum Field Theory and Critical Phenomena,
            (Clarendon Press, Oxford, 1993) 2nd ed.

[Zuk94]   J.A. Zuk,
            *Introduction to the supersymmetry method for the Gaussian
            random-matrix ensembles,*
            preprint, cond-mat/9412060 (1994)

Symbols

Symbol	Description (first occurrence)
$[\cdot,\cdot]$	commutator (Chap. 1)
$\{\cdot,\cdot\}$	anticommutator(Chap. 1)
$[\![z]\!]$	largest integer strictly less than z (Sect. 8.1)
$\mathbf{1}$	unit matrix, unit operator (Chap. 1)
∇	nabla operator (Sect. 2.1.1)
$\sim$	asymptotic behavior (Sect. 3.3.3)
$\approx$	weak equality (Sect. 4.5.1)
$\simeq$	approximation (Sect. 6.1.1)
a	$a(x) = \arcsin\left(\Phi(x)/\sqrt{2E}\right)$ (Sect. 4.4)
a	$a(x) = \arcsin\left(\Phi(x)/\sqrt{E}\right)$ (Sect. 6.1.2)
$a_i,\ a_i^\dagger$	standard bosonic annihilation, creation operators (Sect. 2.1.3)
a_i	components of the vector potential (Sect. 8.1)
a_s	set of parameters (Sect. 5.2)
$A,\ A^\dagger$	generalized annihilation, creation operators (Sect. 2.2.1)
$\mathbf{A}$	vector potential (Sect. 2.1.1)
$b_i,\ b_i^\dagger$	fermionic annihilation, creation operators (Sect. 2.1.3)
$\mathbf{B}, B$	magnetic field, magnetic field strength (Sect. 2.1.1)
c	speed of light (Sect. 2.1.1)
C	normalization constant (Sect. 3.3.1)
$\mathbb{C}$	complex numbers (Sect. 2.1.1)
D	diffusion constant (Sect. 7.1)
e	electric charge, elementary charge (Sect. 2.1.1)
dim	dimension of a space (Sect. 2.2.4)
deg	degree of a Grassmann number (Sect. 4.5.1)
det	determinant of a matrix or operator (Sect. 4.5.2)
$E,\ E_n$	energy, energy eigenvalues (Sect. 2.2.2)
E_0	ground-state energy (Sect. 2.1)

$E_{\text{qc-SUSY}}$	quasi-classical SUSY approximation to energy eigenvalue (Sect. 6.4.2)
$E_{\text{WKB}\pm}$	WKB approximation to energy eigenvalue (Sect. 6.4.2)
$\mathcal{E}$	Grassmann-valued classical energy (Sect. 4.3)
F	Grassmann part of energy (Sect. 4.3)
F	mapping of a set of parameters (Sect. 5.2)
F	magnetic flux (Sect. 8.1)
$\mathcal{F}$	fermion-number operator (Sect. 2.2.1)
g	Landé or gyromagnetic factor (Sect. 2.1.1)
$\hbar$	Planck's constant divided by 2π (Sect. 2.1.1)
H	Hamilton operator, SUSY Hamiltonian (Sect. 2.1)
$H_\pm$	SUSY partner Hamiltonians (Sect. 2.1.2)
H_{loop}	loop correction (Sect. 3.1)
H_s	hierarchy of isospectral Hamiltonians (Sect. 5.2)
H_{tree}	tree Hamiltonian (Sect. 3.1)
$H_{\text{D}}, H_{\text{P}}$	Dirac Hamiltonian, Pauli Hamiltonian (Sect. 2.1.1)
$H_{\text{P}}^{(2)}, H_{\text{P}}^{(3)}$	Pauli Hamiltonian in 2, 3 space dimension (Sect. 8.1)
H_V	standard Schrödinger Hamiltonian for potential V (Sect. 5.1)
H_{T}	total Hamiltonian (Sect. 4.5.1)
$\mathcal{H}$	abstract Hilbert space (Sect. 2.1)
$\mathcal{H}^\pm$	subspaces of positive/negative Witten parity, (Sect. 2.2.1)
i	imaginary unit, $\text{i}^2 = -1$ (Chap. 1)
ind	Fredholm index of an operator (Sect. 2.2.4)
inf	infimum (Sect. 2.1)
k_{F}	Fermi momentum (Sect. 8.2.2)
ker	kernel of an operator (Sect. 2.2.1)
L	Lagrangian (Sect. 4.2)
$L_{\text{qc}}, \widetilde{L}_{\text{qc}}^\pm$	quasi-classical Lagrangians (Sect. 4.3)
$L^2(\mathbb{R}^d)$	Hilbert space of square integrable functions on $\mathbb{R}^d$ (Sect. 2.1.1)
m	mass (Sect. 2.1.1)
$m_t, m_t^\pm$	transition probability density (Sect. 7.1)
M	magnetization (Sect. 8.2)
$\mathcal{M}$	configuration space of Witten's model (Sect. 3.1)
$n_\pm$	number of zero modes of $H_\pm$ (Sect. 2.2.4)
N	number of self-adjoint supercharges (Sect. 2.1)
$\mathcal{N}_\pm$	number of electrons with spin up/down (Sect. 8.2)
N_0	natural numbers including zero (Sect. 3.5.2)
p, p	momentum variable or operator (Sect. 2.1.1)
p_E	magnitude of classical momentum for given energy E (Sect. 6.1.1)

p_E^{qc}	magnitude of quasi-classical momentum for given energy E (Sect. 6.1.2)
P	stationary distribution (Sect. 7.2)
$P^{\pm}$	projectors onto $\mathcal{H}^{\pm}$ (Sect. 2.2.1)
Q_i	self-adjoint supercharges (Sect. 2.1)
$Q, Q^{\dagger}$	complex supercharge (Chap. 1, Sect. 2.2)
q	Grassmann part of classical path (Sect. 4.3)
q_L, q_R	classical left/right turning point (Sect. 6.2)
r, r	position variable or operator (Sect. 2.1.1)
R	residual part of shape-invariance condition (Sect. 5.2)
$\mathbb{R}, \mathbb{R}^d$	real numbers, d-dimensional Euclidean space (Sect. 2.1.1)
Res	residuum (Sect. 6.3)
$S, S^{\pm}$	action functional (Sect. 6.1.1, Sect. 6.1.2)
S_{tree}	action for quasi-classical paths (Sect. 6.1.2)
sgn	sign function (Sect. 3.4)
spec	spectrum of an operator (Sect. 2.1)
t	time (Sect. 2.1)
$T_E,$	period of classical motion with energy E (Sect. 6.1.1)
$T_E^{qc},$	period of quasi-classical motion with energy E (Sect. 6.3)
Tr	trace in $\mathcal{H}$ (Sect. 2.2.4)
$Tr_{\pm}$	trace in $\mathcal{H}^{\pm}$ (Sect. 2.2.4)
U	superpotential (Sect. 3.3.1)
$U_{\pm}$	drift potential (Sect. 7.2)
v	velocity operator (Sect. 8.1)
$V_{\pm}$	full potential in Witten's model (Sect. 3.1)
W	Witten parity (Sect. 2.2.1)
W, W_{tree}	Hamilton's characteristic function (Sect. 6.1.1, Sect. 6.1.2)
x	position variable or operator (Sect. 2.1.2)
x	Grassmann-valued classical path (Sect. 4.2)
x_{qc}	quasi-classical solution or path (Sect. 4.3)
x_L, x_R	quasi-classical left/right turning point (Sect. 4.4)
$\mathbb{Z}$	integers (Sect. 4.5.2)
α, β	Dirac matrices (Sect. 8.3)
β	inverse temperature (Sect. 8.2.1)
Γ	Euler's gamma function (Sect. 6.4.2)
δ	Dirac's delta function (Sect. 3.5.1)
δ_{ij}	Kronecker's delta symbol (Sect. 2.1)
Δ	Witten index (Sect. 2.2.4)
$\Delta, \hat{\Delta}, \bar{\Delta}$	regularized indices (Sect. 2.2.4)

$\varepsilon, \bar{\varepsilon}$	infinitesimal Grassmann numbers (Sect. 4.2)
ε_F	Fermi energy (Sect. 8.2.1)
$\theta, \bar{\theta}$	Grassmann variable, generator of Grassmann algebra (Sect. 4.2)
Θ	unit-step function (Sect. 2.2.4)
λ_1, λ_2	Lagrange multipliers (Sect. 4.5.1)
$\lambda_n, \lambda_n^{\pm}$	decay rates (Sect. 7.2)
Λ	helicity operator (Sect. 8.1)
μ	Morse index (Sect. 6.1.2)
μ_B	Bohr magneton (Sect. 8.2)
ν	Maslov index (Sect. 6.1.2)
$\xi^{\pm}$	canonical momenta of $\widetilde{L}_{qc}^{\pm}$ (Sect. 4.3)
$\xi, \xi^{\pm}$	white noise variables (Sect. 7.1)
$\pi, \bar{\pi}$	classical fermionic momenta (Sect. 4.5)
Π	space-parity operator (Sect. 3.4)
$\sigma_1, \sigma_2, \sigma_3, \sigma_{\pm}$	Pauli matrices (Chap. 1)
$\varphi, \varphi_k^{(i)}$	fermionic phase functional (Sect. 4.3)
$|\phi^{\pm}\rangle$	states in $\mathcal{H}^{\pm}$ (Sect. 2.2.2)
$|\phi_E^{\pm}\rangle$	eigenstates of $H_{\pm}$ in $\mathcal{H}^{\pm}$ (Sect. 2.2.2)
ϕ_0^-	zero-energy eigenfunction of Witten's model (Sect. 3.3.1)
Φ	SUSY potential (Sect. 2.1.2)
$\Phi_{\pm}$	asymptotic values of the SUSY potential (Sect. 3.4)
χ	paramagnetic susceptibility (Sect. 8.2.1)
χ_0	paramagnetic zero-field susceptibility (Sect. 8.2.1)
χ_1, χ_2	second-class constraints (Sect. 4.5.1)
$\chi_{\pm}$	spinors (Sect. 8.3.1)
$\psi, \bar{\psi}$	classical fermionic degrees of freedom (Sect. 4.1)
$\psi_0, \bar{\psi}_0$	generators of a Grassmann algebra (Sect. 4.1)
$|\psi\rangle$	states in $\mathcal{H}$ (Sect. 2.2.1)
$|\psi_0\rangle, |\psi_0^j\rangle$	SUSY ground state(s) (Sect. 2.1)
$|\psi^{\pm}\rangle$	positive/negative Witten-parity states (Sect. 2.2.2)
$|\psi_E^{\pm}\rangle$	energy eigenstates with positive/negative Witten parity (Sect. 2.2.2)
ω	frequency (Sect. 2.1.3)

Name Index

Subject Index

Springer-Verlag
and the Environment

We at Springer-Verlag firmly believe that an international science publisher has a special obligation to the environment, and our corporate policies consistently reflect this conviction.

We also expect our business partners – paper mills, printers, packaging manufacturers, etc. – to commit themselves to using environmentally friendly materials and production processes.

The paper in this book is made from low- or no-chlorine pulp and is acid free, in conformance with international standards for paper permanency.